STAR WARS

WHERE SCIENCE MEETS IMAGINATION

INTRODUCTION BY ANTHONY DANIELS (C-3PO)

PUBLISHED IN ASSOCIATION WITH LUCASBOOKS AND THE MUSEUM OF SCIENCE, BOSTON

NATIONAL GEOGRAPHIC
WASHINGTON, D.C.

TABLE OF CONTENTS

In a war of technologies, the Republic's clone army takes on the Separatists' battle droids during a showdown on Geonosis.

FOREWORD

Ed Rodley

ED RODLEY has been developing exhibitions for the Museum of Science since 1987. He is responsible for research and development, including content research, interactive exhibits, and label writing.

DESPITE THE TITLE OF THE MUSEUM OF SCIENCE EXHIBITION AND THIS BOOK, THE SUBJECT really isn't *Star Wars* per se. It's about our world, and our future. So, why *Star Wars*, then? One of the great challenges of thinking about the future lies in visualizing things, particularly technologies, that don't yet exist. The Museum of Science has been wrestling with ways to address technological literacy in response to a growing concern among educators that we don't do enough to give people the skills to manage their relationship with technology. In 2002, The National Academy of Engineering (NAE) and the National Research Council (NRC) stated that Americans must become better stewards of technological change and that technological literacy is essential for people living in a modern nation like the United States.

When we started brainstorming ways to excite people about imagining the future and taking a more active role in shaping it, we quickly ran into a problem: We have no shared vision of what the future looks like. To talk about our future, we needed a depiction of a technologically advanced society that was familiar to a broad audience. It would also have to be pretty exciting, something that would energize people and make them want to learn more about things that, at first blush, are pretty esoteric. We needed something that also gave us fertile ground to explore the implications of technology decisions. What we needed was *Star Wars*. The attractiveness of the images created in the *Star Wars* movies is not accidental. Led by writer-director George Lucas, the movies' concept artists, model makers, and technicians tapped into universal human needs and dreams in a way that has powerfully captured people's imaginations for almost three decades. Science centers and educators nationwide need to be able to do that as well.

The goal of this book and the exhibition has been to view *Star Wars*, not as a blueprint, but as an inspiration, as a springboard to think about what our future could be like. Here was a way to use *Star Wars* to do something other than talk about moviemaking. The films' depictions of an entire galaxy and its technologies can help us imagine what our own future could be like. As you'll see in many of the essays in this book, scientists and engineers all over the world are doing precisely that. They've seen something fantastic, and thought "Hmm, how could I do something like that?" And the answers to these musings are every bit as exciting as the movies—some would say even more so, because they're real.

Star Wars: Where Science Meets Imagination uses the science fantasy of the *Star Wars* films to inspire us to examine different future possibilities, even if they may seem like impossibilities today. The exhibit—and this book—is organized around two technological challenges that are depicted in the *Star Wars* movies—transportation, and robotics. Visitors to the exhibition can explore how those challenges are addressed in the *Star Wars* universe, using over eighty props and costumes from the Lucasfilm Archives spanning all six films. Movie clips, interviews with the filmmakers, and immersive experiences give visitors a glimpse into the creative process and let them explore the *Star Wars* galaxy themselves.

Real-world artifacts demonstrate how far we have come in designing technologies that increasingly look like they're science fiction. Hands-on interactive exhibits explore real-life 21st-century technological solutions to some of the same challenges the filmmakers faced. Visitors can test-drive a floating vehicle, ride a maglev train, and try to make a robot walk. Airspeeders meet maglev trains, and droids meet real robots. Comparing and contrasting is only the starting point. To really understand any process, you actually have to engage in it. The focal points of each theme area are found in two engineering design laboratories, where visitors can design, build and test their own magnetically levitated speeder and their own robot.

One of the most powerful experiences of working on the exhibition and this book has been meeting some very creative scientists, engineers and filmmakers, and realizing how central a role an active and joyful imagination has played in all their endeavors. It is something we can all share in. And its place in helping us learn—indeed, making us hungry to learn—is central. The educator Herbert Kohl, author of *The Open Classroom* and *Growing Minds*, was speaking of young children when he wrote the following words, but they apply to all of us:

> *What is the place of imagination in education? The word imagination does not appear in the government's...lists of behavioral objectives or educational outcomes. There is no imagination curriculum or pedagogy of the imagination in our schools. Yet if, as the poet Wallace Stevens wrote, "the imagination is the power of the mind over the possibilities of things," then to neglect the imagination is also to impoverish children's worlds and to narrow their hopes.*

Hence the title of the exhibition and this book. Enjoy!

SEE-THREEPIO AND I

Anthony Daniels

Anthony Daniels left the study of law to pursue a career in television, film, and on stage; he appears in all six *Star Wars* films.

C-3PO APPEARED IN THE DESERT, JUST AS ONE OF THE TWIN SUNS FLAMED ON THE HORIZON. He—or I—or, he *and* I gazed at the world for the first time, and the world stared back in wonder.

We were all amazed.

In 1975, my initial reaction to the offer of a meeting with George Lucas had been a question. Why would I, a serious actor, even consider discussing a role in a low-budget science fiction film, especially the role of a robot?

Perhaps I'd been traumatized in 1953 by weekly installments of *Quatermass*, the scary, science fiction drama that I first watched as a seven-year-old, peeping out at the tiny black-and-white TV screen from behind the safety of the sofa. And, in my book, robot could almost be spelled D-A-L-E-K, the villains of the original *Doctor Who* series I had quite enjoyed some years later. Certainly it had never occurred to me that I should personally make a living inside a box, wheeling around and waving a sink plunger but never using the stairs. Robots, Daleks, sci-fi—they had nothing to do with me. I had just spent three years at acting school learning to enunciate. And always as a human.

There was nothing of Hollywood or showbiz about my first meeting with George. We had a rather tepid conversation about my skills as an actor and especially my abilities in mime. He knew that the person inhabiting the sort of costume he had in mind would have to possess a great deal of physical self-control and expression in order to create a believable character from inside an unexpressive suit of armor. But I was only half listening as he outlined the rest of his project. I had noticed a painting of the golden robot on his office wall.

Ralph McQuarrie was the production illustrator. He drew airplanes for Boeing in a previous incarnation, and had clearly made a huge artistic leap. Here he had painted two mechanisms in an alien moonscape, apparently bereft of human contact. And yet I saw great humanity in that picture. The forlorn expression and penetrating gaze of the taller humanoid spoke to me as clearly as some people find meaning in the Mona Lisa's quirky smile.

In any event, I got the job.

My translation into the outward appearance of Threepio began with a rather peculiar casting session. No illicit couch—just buckets of plaster, flung at my body to create a mold

The protocol droid C-3PO embodies a classic irony: He speaks countless languages but can never quite understand the motives of the humans around him.

on which the now iconic shape would take, well, shape. The sculptor, Liz Moore, applied clay to the cast of my body, quickly disguising the human curves and angles. Then the props and engineering department took on the task of manufacturing mechanical ankles, elbows, shoulders, knees, and all the bits in between, so that George's vision could come to life, or, at least, be switched on.

Replicating Nature was not easy.

It had all taken six months, and now the day had arrived.

From inside Threepio's head, my new vantage point—or as I would soon realize, disadvantage point—I could see far across the desert to the horizon's curve. On the other hand, I could see only what was straight ahead of me. To the sides, I was as blinkered as a fairground donkey. But I still sensed the sun's early rays making the costume glow. As I slowly turned my head, I saw the American team standing to my left, wide-mouthed with loud expressions of astonishment. Strewn in front of me, the British crew nodded quietly in respectful appreciation. The local Tunisian helpers, clustered to my right, almost fell to the ground in awe of what may have seemed like the second coming, or at the very least a close encounter of the hitherto unknown kind.

Threepio was a hit. A star. A wonder.

Then the costume broke. The props department patched me up and the cameras rolled, somehow avoiding my human foot sticking out of Threepio's damaged metal shin. The illusion was preserved...until the robot's eyes shorted out and off, which took longer to fix. Increasingly frustrated, George frowned at his creation and I looked back, feeling it was all my fault.

If Threepio was a shock for George, R2-D2 was my surprise.

I had studied the scripted words of our close relationship for months, and finally the time came for us to meet. It wasn't propitious. As the remote-controlled blue and white box rolled up to me, right after the Jawas had sold us to Luke's uncle, Owen Lars, I said, "Now, don't you forget this! Why I should stick my neck out for you is quite beyond my capacity!" I turned and led the way into the Homestead, or, rather, to the stairs leading down to the underground interior. Like the Daleks, Threepio didn't do stairs. (The suit's brilliant mechanical design couldn't provide the sort of gyroscopic balance we would see years later in Honda's wondrous android, ASIMO.)

George yelled, "Cut!" from behind a distant camera, and I stood still at my stop mark at the top of the stairs. Proceeding even just a few inches farther would have caused severe damage to me and, more importantly I suspected, to the costume.

I stopped. Artoo didn't.

Fifty pounds of mechanics, radios, and batteries seemed intent on wounding me and wrecking itself. The crew ran to find Artoo's off switch while I briefly pondered the laws of robotics, especially the one about robots not harming themselves or a human. Here was a robot trying to do both.

Gleaming like an idol in the sun, C-3PO has impressive programming but is often taken for granted by the humans he serves.

The second take went no better. If anything, Artoo's overcharged batteries seemed to give added power to an apparent desire to plough me over. George, his patience clearly stretched, muttered comments to the crew operating Artoo's radio remote. George's serenity was clearly as strained as my nerves. On take three, I repeated my line and action as before. With one change: At the last minute, I courteously stepped aside and ushered Artoo ahead of me. On the brink of disaster, the mechanized box stopped, saving itself from the fate it had just wished on me. I felt perhaps I should be concerned about my counterpart's loyalties.

I had had months to learn the script, which always mentioned Artoo's replies as beeped observations, expletives, and responses to my own lines. In reality there was silence, apart from the odd whirring of Artoo's motors. I soon learned to fill in the gaps with lines I imagined for my co-star. I spent weeks pretending to bond with a box, having conversations with a box, caring about a box. People who do that sort of thing are usually given medication. For an actor, life with a box is better than no life at all—but not much.

Of course, Artoo's lines would be added months later by sound designer Ben Burtt. The electronic beeps that would help Ben win his first Oscar were created on the keyboard of an antique plastic synthesiser. Magic. But the real imaginative genius of his creation came from the addition of his own voice among the electronics. His human whistles, sighs, and moans give the audience a feeling of familiarity with the blue-domed machine. They sense an organic life form that they recognise as part of themselves. They anthropomorphise, as they would with a pet animal, finding all sorts of human qualities in something that is, in this case, mere metal and plastic. And George's superb script provided the basis for the "odd couple" double act that I was now busily creating, alone.

Threepio would wish to remind you that Artoo is only an astromech: a robot that may have the mechanical genius of a motorized Swiss Army knife but the sensitivity of a drill sergeant, whereas Threepio firmly grasps the esoteric skills of protocol and etiquette, so essential for the smooth functioning of society on our planet—though utterly useless in a hectic space adventure played out on Tatooine.

Unlike the clearly utilitarian Artoo, See-Threepio is a humanoid, socialised robot with extraordinary skills, most of which are ignored by the humans he so diligently tries to serve. The design and manufacture of any appliance reflects its purpose. Imagine the frustration level of a machine whose function is repeatedly denied. No wonder Threepio seems a little tense: His outstanding linguistic abilities are regularly taken for granted. Capable of advanced mathematics, he must endure the routine dismissal of his numeric prognostications by the ill-mannered Han Solo, a human who has no respect whatsoever for the golden droid—a droid that, amazingly, appears to have a soul, which occasionally peers through the mayhem of the *Star Wars* saga.

The saddest omission for Threepio and me occurred in *Attack of the Clones,* when we filmed a scene between Padmé and the unclothed See-Threepio, without his metal coverings. The

two had last met many years earlier. Unable to sleep, the young woman discovers the droid quietly sitting, alone, in the garage on Tatooine—his wiring and motors still exposed, as on the first day of his creation. She asks if he is happy there, and the sensitive machine replies that he is not un-happy; it's just, well, to be like that....Like what? Naked! If you'll pardon the expression. It simply wasn't protocol. And for a machine whose primary function is etiquette, to be unclothed in the world...well! Exquisite, enduring humiliation.

I found it one of the most tender moments in all six films, a chance for the audience to glimpse whatever area of electromechanics might harbour Threepio's soul. But, sadly, the scene was later deleted from the final cut.

Threepio's unusual programming gives him the role of facilitator between machines and humans, leading him to recognise his similarity to the latter. It also makes him aware that he just lacks that final *something*, that synaptic spark, which would make him human and qualify him to be treated like a human, too. His tragedy lies in his inherent inability to make that leap to instinctive emotion and true humanity. He is forever doomed to remain a robot. Given his depth of understanding, that thought might well bring a tear to his photoreceptors, were it not that his creator failed to install that particular program!

It's not easy being gold: Anthony Daniels requires assistance to suit up for a photo shoot as C-3PO. Daniels has made his career playing the anxious droid in search of respect.

I learned the true place of the robot in our world during the first days of shooting the 1977 *Star Wars*. Actors in films usually receive the gentle ministrations of the wardrobe staff. I was assigned to Maxi from the props department. On Day One in the North African desert, Threepio was greeted with open-mouthed astonishment. They came up and told me how wonderful I looked. They praised my performance. They took souvenir photos. But by Day Two, they had accepted the extraordinary robot as—a robot. A machine without feelings. An object they could take for granted and ignore, mirroring the script's insight into human attitudes, leading our metallic hero to sadly note, "Nobody worries about upsetting a droid. We seem to be made to suffer. It's our lot in life."

In one day, I had stopped being an actor. I was a thing.

Years later, when I met ASIMO, a robot designed to help the elderly and the ill, I was shocked at the ease with which the designers referred to him as "It." Much as I wanted to acknowledge its human quality, they rationally denied it. Yet they created it in a human form precisely so that those it serves can relate to it. Personally, I'm going to have to think more about that. How human-like do we want our mechanical companions to be? How human can they become before gender is an issue? On a visit to the Lego factory in Denmark, I was fascinated by the witty-looking robots that stamped out multicolored blocks of plastic, and by the mobile units that collected the finished pieces and delivered fresh supplies to their counterparts. They worked in simple harmony and what appeared to be joyful collaboration. On leaving, I was shocked when my host turned out the lights and left them to trundle their way in the dark.

I'm not alone in searching out the human qualities of machines that work hard to make human lives easier. How many cars, boats, planes are referred to as "she?" At times, it seems we all need and want to anthropomorphise in order to make life feel more comfortable. It's fun, too. My last shot in Episode III was a scene between Artoo and Threepio. I walked along a blue wall on a blue carpet. Artoo, George explained, would be digitally added later by ILM. So could I please talk to myself as usual? Well, after twenty-eight years that wasn't going to be a problem. Then I noticed the short, domed vacuum cleaner they had just used to clean the carpet. For fun, in rehearsal, I tugged it along beside me as I talked to it. The crew laughed, although the impromptu scene failed to make the final cut of the movie. But knowing Threepio so well has possibly made me even more thoughtful as to how a machine—a robot—might feel. And how I should react.

Increasingly, we're structuring our lives around screen-based interactions. Keyboard skills are flourishing, but what about social skills? Future generations may well find it easier to relate to machines than to their fellow humans. The binary route isn't two people getting together in reality. It is communication reduced to zeros and ones in a computer's innards, filtered across the ether. Could the increase in technological interaction foreshadow the end of real human relationships as we know them? Sounds like sci-fi to me.

The first *Star Wars* film seemed to place droids in close harmony with man, featuring

Threepio as a useful, socialised household object. Later movies in the saga contain battalions of vicious machines designed with a more deadly application. It happens in the movies. It's going to happen on Earth, where even nonaggressive, domestic machines can intimidate humans. Yet, whether we like it or not, whether we understand it or don't, we are all going to rely more and more on technology in the future—a technology managed by humans. It is a human brain that inputs the codes and creates the algorithms. It is the silicon brain that acts out the program. But always there is that human seed which started the whole process. To create a truly autonomous, cognitive robot would require a loosening of the restraints: a willingness to let a machine do its own thing, to grow and learn, to let it out into the world.

I imagine George has created something of a problem for technologists. In movies it's very easy to make things look very easy. Yes, Threepio and Artoo have all sorts of astonishing skills and physical abilities, but only through the power of editing and postproduction. I've had to explain even to seasoned film crews on spin-off programs that, for instance, Threepio can't sit down on camera: You see him standing; you insert a cut-away shot, then move back to him settling in a seat, or sometimes, on a throne. The bits in between constitute a messy interlude of piecing together a camera's-eye-only version of a sitting costume. But you never get to see that. To the audience the moment appears seamless.

So when the worlds of *Star Wars* show off all sorts of inventive devices and gadgets, the expectation factor on planet Earth is huge. The public can be thrilled by the reality of scientific invention, but sometimes disappointed too. Why can't it be like it is in the movies? But as you will read in this book, understanding and technology are catching up.

When we were making the original *Star Wars*, video recorders weren't even on sale to the public. As we made the last film in the saga, they were becoming obsolete landfill. That's progress and it's getting faster. Personally, I never knew how to program the thing anyway. I may play a brilliant machine but I wouldn't necessarily know how to use one. I know what you're thinking—read the instruction manual. But hey, I'm only human. Still, I am happy if Threepio has contributed to an appreciation of the supportive role that robots and technology in general will need to play in the human adventure, wherever that might be leading us.

What intimidates me are the real robotics geniuses in the world: scientists and robotics engineers who have often been inspired by the freewheeling imagination of artists and dreamers like George Lucas. Their thoughts in this book make fascinating reading. And while I feel somewhat unworthy in the face of their growing fund of knowledge, might I just point out that, unlike me, they really don't know what it's like to be a machine. I can tell them. It's tough. So please, be nice. I've been known to say a friendly "good night" to the cherry-red blink of the alarm system as I climb the stairs to bed. It always winks back. It gets the message. And please, be appreciative. I admit I sometimes get tetchy with my computer, but on other occasions I have spoken to it with lavish gratitude. Not, of course, if anyone else is in the room.

George has always blurred the edges of man and machine in his *Star Wars* saga. A strange preoccupation with the various dismemberments of Threepio is mirrored by the amputations inflicted on assorted heroes and villains, whose limbs get rapidly replaced through the wonders of medico-mechanical engineering. It happens in the movies. It's happening on Earth. It could help to preserve us all.

If I have a concern, it is this. As I write, I believe that Threepio is the only bipedal, free-standing, self-energizing, cognitive, sensitive, autonomous robot in the galaxy. Without wishing to be a Luddite, a destroyer of innovative technology, I'd like to beg a favour of all those scientists and engineers whose work is advancing at a rapid and exciting rate: Please, slow down just a bit, before you make me redundant. I need the work!

As you may have guessed, I have developed a fondness for my slightly eccentric friend over the nearly 30 years we have been counterparts. I know his talents will never be taken for granted by humans of imagination and intelligence. On his behalf as well as my own, I am grateful for the friendship and acceptance audiences have shown him. And it is only right to thank the Maker, George Lucas, without whom none of us would ever have known a golden robot called C-3PO.

Chief droid wrangler Don Bies adds the final polish to C-3PO's famous golden shell. Anthony Daniels has played the same character in all six *Star Wars* films, over a span of nearly 30 years.

PART ONE

GETTING AROUND

The Future of Transportation

lot of ship for

THE FUTURE OF TRANSPORTATION

ED RODLEY

GETTING FROM PLACE TO PLACE IS ESSENTIAL FOR ALL OF US. WE NEED TO GET FROM home to work, or school, or the store, and we rely more and more on sophisticated technologies to do so. In the *Star Wars* galaxy, people zip around on speeders to travel short distances—say, from ground to orbit—while faster-than-light starships carry people and goods from planet to planet. Five hundred years ago on Earth, most people never traveled farther than they could walk on their own two feet. Now, we routinely cross continents at 500 miles per hour for business and just for fun.

Our technology has evolved. Computers control our cars, sophisticated networks of people and computers monitor automobile and air traffic and keep trains and planes on track and (we hope) on time. Whatever your choice of transportation technology, be it a bike, car, or helicopter, it is part of a largely invisible infrastructure of control systems and social constructs, like laws, created over decades and constantly struggling to adapt to changes in technology. The side of the street you drive on has nothing to do with physics. It is a decision that was made by people in response to a particular historical situation. In fact, it isn't much of an overstatement to say that transportation networks have determined how our cities and towns look. A modern Western city devotes as much ground space to accommodating cars as it does people. And as we venture farther—into orbit, to the moon, and beyond—the means we use to get there will shape us as much as we shape them.

What will the coming century's transportation technologies do to the fabric of our lives? Dean Kamen, the inventor of the Segway, uses the example of the horseless carriage. When automobiles first became available, they entered a world ruled by horses. Horse-drawn wagons, carriages, and buses filled the streets. There was no obvious way of accommodating these automobiles, so they were shoehorned in with the existing modes of transport and expected to follow seemingly normal rules, such as having someone walk in front of every car with red flags to warn other traffic of an automobile's approach. It was only when the automobile took off that people began to think that maybe the old rules might have to change. People's changing attitudes are what made the car all-powerful, and so it will be for whatever replaces the automobile.

A possible key to interstellar travel, antimatter engines will be designed to harness the enormous energy of matter/antimatter collisions—ten billion times the energy of chemical reactions. Europe's CERN laboratory uses giant magnets, like the prototype at left, to study and produce antimatter.

For the generations of moviegoers who saw the original *Star Wars* trilogy in the 1970s and 80s, some of the most enduring images are the spacecraft: the first time an Imperial Star Destroyer rumbled overhead for what seemed like minutes, the rattletrap *Millennium Falcon*, the angry snarl of TIE fighters going by. Part of the appeal of the movies could be found in the breadth and depth of George Lucas's vision of the *Star Wars* galaxy and the way that the characters interacted with their technology.

The Empire had shiny new ships and the ability to create technological terrors on a scale never before imagined. The Rebel Alliance seemed to make do with ships that hadn't seen a coat of paint in years, looking and sounding as though they were held together only through constant effort. And at the far end of the spectrum, there was Han Solo, greasy rag in hand, laboring over his recalcitrant, cantankerous freighter, souped up almost to the point of being nonfunctional. Here was a vision of futuristic technology that reminded us of our own world. Things got dirty; things broke. Some people had the latest equipment; others had to make do with hand-me-downs. It all fit together to make a perfectly plausible universe. And it tantalized us with the notion that new worlds were only a jump away.

In the thirty-odd years since *Star Wars: A New Hope* came out, a lot has happened in space travel. The so-called space race between the United States and the Soviet Union is long over. The latest in a series of international space stations sits in orbit, slowly taking shape. The number of astronauts, cosmonauts, and taikonauts who've entered space is in the hundreds. We've seen space tourists, and now the first commercial space vehicle has gone into orbit. Plans are already afoot for the first regular passenger service into low orbit. A lot has happened, but we're still largely confined to our one little planet. When, if ever, will we be able to hop in our spaceship and explore the distant planets of our universe, or beyond?

The answer depends on our sense of time and scale. We have already begun exploring planets millions of miles from Earth using robotic explorers. In the last few years, NASA has revived plans for returning to the moon and sending people to Mars. If the solar system isn't exotic enough for you, there are a number of technologies being developed to send ships into the completely untraveled gulfs of interstellar space.

Once you start talking about moving between stars, you bump into speed limits—specifically, the speed of light. Our understanding of the universe does not allow us to exceed it, so zipping to Alpha Centauri and back is not an option. But, as Lawrence Krauss points out, if you don't mind taking a few years to get there, a number of potential technologies exist that you might employ. Using these technologies, you could build a spacecraft that could take you to one of the stars in our local neighborhood in less than a lifetime. All you'd need are trillions of dollars and more materials than humanity has ever assembled for one purpose. It wouldn't be easy, but that's not the same as impossible. We haven't left our solar system yet, partly because we haven't decided to.

On a more terrestrial level, *Star Wars* presents us with a vision of how we might interact with our technologies. Both trilogies start out with a boy and his car. Luke's landspeeder and Anakin's homemade Podracer represent an entire class of frictionless ground vehicles that have captivated generations of *Star Wars* fans. Obvious parallels exist between them and the hot-rod culture that George Lucas grew up with in Northern California in the 1950s and 60s, and immortalized in *American Graffiti.* They are hand-built, customized vehicles that their owners know inside and out, and can easily troubleshoot and repair. How many of us can do that with our cars today?

In the 1970s, a pessimistic view of technology and the future dominated American popular culture. Technology was something to be feared, something uncontrollable. In the *Star Wars* universe, though, people seem to understand their technology. The heroes' high-tech vehicles were old, last-year's models, and in the case of Luke's speeder, not worth much. George portrayed a vision of our relationship with technology that reflected the audience's desire, conscious or not, to understand their own technological world. But how do we get from here to there?

Marc Millis presents some wonderful thought experiments using scenes from *Star Wars: A New Hope* that perfectly illustrate the power of a compelling image. In the *Star Wars* galaxy, all kinds of vehicles like speeders and Podracers use an anti-gravity "repulsorlift" technology to speed above the ground and even fly into space. The qualities that repulsorlift vehicles possess are impressive. Comparing them to real-world vehicles that don't touch the ground, like aircraft, hovercraft, and magnetically levitated (maglev) vehicles, lets us explore how things work in the real world.

Take any of the scenes of multitiered networks of traffic filling the skies of Coruscant and try to place them on Earth. Imagine a futuristic city where everyone drives vehicles that combine the ease of use of a car with the flight capabilities of a helicopter. What would that city look like? You not only have to subtract the cars from your mental image, but all the support structures that make cars viable—streets, parking spaces and lots, gas stations, traffic lights. Now put it all back together again with new infrastructure. Where do you store flying cars? How do you determine right-of-way in 3-D space? Do *you* even do it, or does a computer drive for you? If all the traffic is up in the sky, would that make ground-floor dwellings more stylish and attractive? If you could go straight from point A to point B, would our cities grow more vertical? Would you want a sixteen-year-old driving a ton of metal over your house at high speeds?

These questions have ramifications far beyond the development of the technology itself, and in fact most of them are completely beyond the control of any inventor to answer. As Hiroshi Nakashima and Sam Gurol both illustrate in their work, the floating maglev train isn't far away. What else is on the horizon and how will we decide whether we should adopt it? These are the kinds of questions that will have to be addressed by you and me, the citizenry. Let's hope we'll be up to the challenge.

SPACE, TIME, AND STAR WARS

LAWRENCE M. KRAUSS

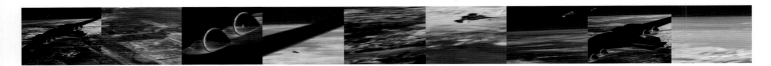

LAWRENCE M. KRAUSS, Ambrose Swasey Professor of Physics, Professor of Astronomy, and Director of the Center for Education and Research in Cosmology and Astrophysics, Case Western Reserve University. The author of numerous publications, Prof. Krauss researches such topics as elementary particle physics, general relativity, and the nature of dark matter.

IN A GALAXY FAR, FAR AWAY, THINGS MAY WELL EXIST THAT WE MAY NEVER BE ABLE TO SEE, precisely because of the same strange features of general relativity that might also in principle allow such exotic phenomenon as faster-than-light space travel—one of the staples of the *Star Wars* galaxy. Current knowledge indicates that the expansion of our universe, first discovered by Edwin Hubble in 1929, is in fact speeding up, so that over time distant galaxies are not only receding from us, but their recession speed is increasing with time—in some cases, at relative velocities that exceed the speed of light!

Those of you who remember from high school that objects cannot move through space faster than light-speed are correct. However, space itself, according to what we now understand about general relativity, can do whatever the heck it wants. Space can expand faster than light-speed, carrying distant objects—which themselves may be at rest in space—away from us faster than light, just as a current may carry swimmers to shore at a high speed, even though the swimmers may be at rest, and not actually swimming at all. Or worse, it may carry the swimmers away, faster than they can ever swim back....

So, if distant space is expanding away from us at an ever increasing rate, galaxies carried along in space may be moving away from us faster than the speed of light, even if they are at rest relative to their nearby surroundings. And the light they emit, which travels through that nearby space at light-speed, cannot overcome the expansion rate of the vast region of space between us and them, so that this light will never, ever be able to cross the vast empty chasm that separates us. Like the unfortunate swimmers, it may disappear forever.

I bring up this eerie, and still puzzling, feature of the universe, discovered only within the past decade, because it illustrates a central fact to remember in comparing the real world and the world of science fiction: Truth is stranger than fiction, and the real universe is far more exotic than anything science fiction writers are ever likely to create. But we can nevertheless have good fun, at the very least, in considering what aspects of the *Star Wars* universe might bear some resemblance to the real world. From a short-term practical perspective, I'm sorry to say, things don't look good. However, in a longer view, the recent discovery regarding the universe's rate of expansion vindicates, at least in principle, the most famous line in *Star Wars*—"in a galaxy far, far away."

At the Mount Wilson Observatory in California, Edwin Hubble discovered that spiral nebulae are not objects within the Milky Way, but rather separate galaxies entirely outside our own, which move away from us as our universe constantly expands.

We, who appear to be stuck to the Earth like glue, yearn to travel to the stars. We invent imaginary technologies that will propel us at light-speed or faster, or we imagine visitors from afar who we hope have come in peace.

Whenever I state in public that I expect we will *never* travel on round-trip voyages to distant stars or galaxies, it provokes debate, but the reason for my pessimism may come as a surprise. What will most likely determine the success or failure of interstellar travel will not be physics we have yet to understand, but physics we have understood for at least 100 years.

It is often said that money is unimportant if one doesn't have one's health. However, both money and health issues provide formidable barriers to successful long-term space travel. In the case of money, I am talking here about fuel. And the amount of fuel needed on board a spacecraft to propel it to near light-speed is almost unfathomably large. For example, if conventional rocket fuel were used, the mass of fuel required to accelerate a spacecraft to, say, half the speed of light would be greater than the mass of our entire galaxy!

We don't need anything as fancy as general relativity or quantum mechanics to understand this dilemma. It remains a simple consequence of Newton's Laws which prove valid for baseballs and for spacecraft—at least until their speed gets close to light-speed—that it is extremely costly to propel spacecraft to speeds much faster than the speed at which the fuel leaves the exhaust. In the case of conventional fuel, exhaust speeds are less than 5 kilometers per second, while the escape velocity from Earth is about 11 kilometers per second.

At the same time that we explore new propulsion technologies, another impediment to space travel arises: that all-important matter of health. Space is permeated by high-energy cosmic rays emerging from objects ranging from our own sun to those galaxies far, far away. We on Earth receive protection from much of this potentially harmful radiation, thanks to our atmosphere and to Earth's magnetic field, which tends to deflect high energy charged particles on trajectories toward the poles. In a spacecraft, unprotected by these defenses, astronauts would be exposed to potentially lethal doses of radiation on round-trips as short as a year or two. Shielding against this hazard would require either thick shells of material surrounding the craft, or strong magnets that could deflect ionizing radiation. However, these fixes add mass to the spacecraft, and mass remains the great enemy since the amount of fuel required increases in direct proportion to the mass of the ship.

Which brings us right back to fuel. New sorts of propulsion methodologies, like ion drives and nuclear electric engines, are being examined as we explore ways to achieve velocities in excess of 30 kilometers per second—speeds required to travel to the outer planets in periods on the order of years—but even these technologies will not allow us to achieve the speeds necessary to imagine round-trips to even nearby stellar systems. At current rocket speeds, it would take over 100,000 years to travel to the nearest star.

We can imagine technologies, such as nuclear fusion, that would produce energetic exhaust streams traveling at perhaps one percent of the speed of light. But even if we could control nuclear fusion, the process that powers the sun and thermonuclear weapons (which

Newton's Laws of Motion: Formulated by Sir Isaac Newton, these 3 laws state that a body at rest or in motion will maintain that state if not acted upon by an outside force; a change in motion can only be caused by another material body; every force causes an opposite and equal reaction.

Thermonuclear: Relates to the transformations in the nucleus of low-atomic-weight atoms, like hydrogen, that require a very high temperature for their inception.

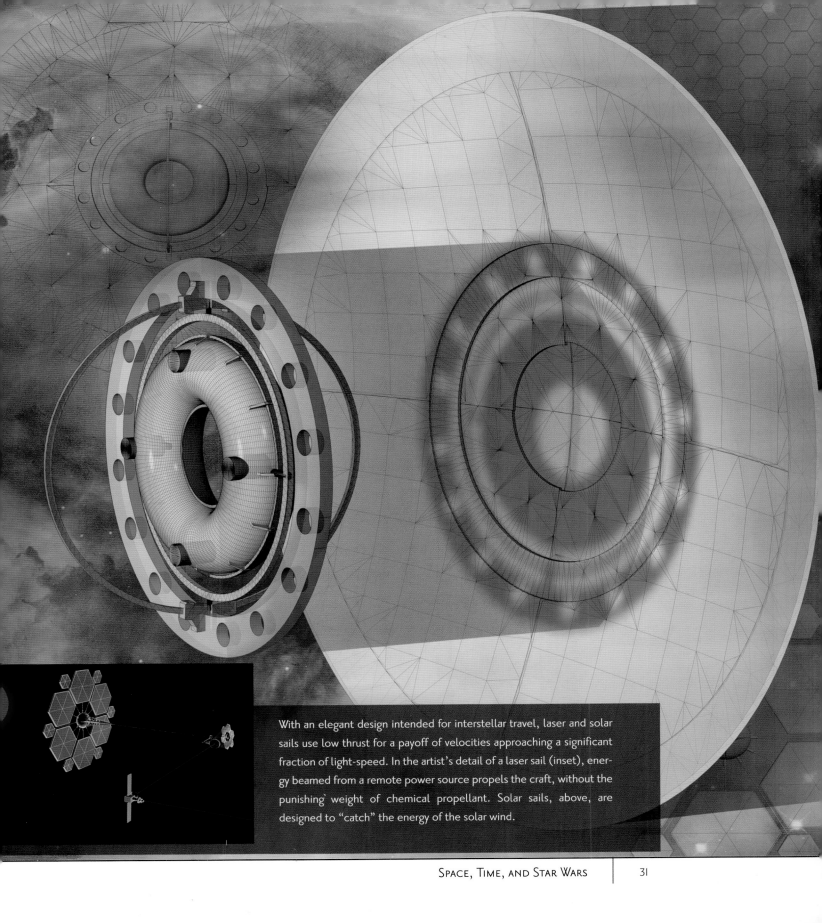

With an elegant design intended for interstellar travel, laser and solar sails use low thrust for a payoff of velocities approaching a significant fraction of light-speed. In the artist's detail of a laser sail (inset), energy beamed from a remote power source propels the craft, without the punishing weight of chemical propellant. Solar sails, above, are designed to "catch" the energy of the solar wind.

In *Star Wars*, ships like the *Millennium Falcon* use fictional technology to travel vast distances at light-speed and beyond. While exciting for moviegoers, the jump to light-speed presents real-world problems. For instance, temporal relationships will change as objects approach light-speed, causing interstellar travelers to age at a much slower rate than humans residing at either the origin or the destination planet.

we cannot at present), the amount of fuel required to accelerate to half the speed of light and then stop would be in excess of 7,000 times the mass of the spacecraft we wanted to propel!

Perhaps the only possibility for realistically accelerating spacecraft to relativistic speeds lies in using fuel from outside a spacecraft. A powerful laser system could potentially impart sufficient momentum to a spacecraft endowed with a huge solar sail to reflect the light even as it moves a large distance away, accelerating to perhaps a tenth the speed of light over several years. Of course, the solar sail required would be nearly the size of Texas, and this mechanism does not easily allow for deceleration. Clearly, using propulsion to send humans on round-trips outside of our solar system is best left to the movies. For realistic travel to the stars, we'll have to consider other alternatives.

MANIPULATING SPACE OR MANIPULATING TECHNOLOGY?

Einstein's theory of general relativity revealed the remarkable fact that both space and time can dynamically respond to the presence of matter and energy. Space can curve, expand, and contract, for example. Indeed, the expansion of our own universe provides explicit testimony to the remarkable features of gravity as Einstein explained them to us.

Once we allow for the dynamic response of space to matter and energy, a host of new phenomena become possible, at least in principle. Among these are phenomena reminiscent of *Star Wars'* hyperspace, or *Star Trek's* warp drive. If we could manipulate space appropriately, then movement from one location to another would not require propulsion of a craft *through* space, but rather propulsion of space itself!

As I've already said, the laws of physics that imply one cannot travel through space faster than the speed of light impose no such limits, at least as far as we know, on the expansion of space itself. Indeed, there are currently distant objects in our universe that are likely receding from us faster than the speed of light—objects that are, in their local surroundings, not moving at all. The very space they inhabit carries them and their nearby neighbors away from us. Could we then somehow learn to manipulate space, using gravity, so that space behind a craft could be made to expand rapidly while space in front of the craft quickly collapsed, enabling the craft to "jump to hyperspace," like the *Millennium Falcon?*

To do so would require generating a kind of energy, called negative energy, which we thus far have no idea how to create and maintain in the laboratory. It is quite plausible, in fact, that when we understand how to combine general relativity and quantum mechanics better, we will be able to prove that such a scenario is impossible. We simply do not know at the present time. What we do know is that even if we could generate this exotic form of energy, the actual magnitude required to produce macroscopic effects would likely be immense. Early estimates, for example, of the amount of energy needed to propel a reasonably sized vessel at "warp speed" exceeded the total energy available in our entire galaxy. Once again, the prospects for light-speed travel do not appear rosy.

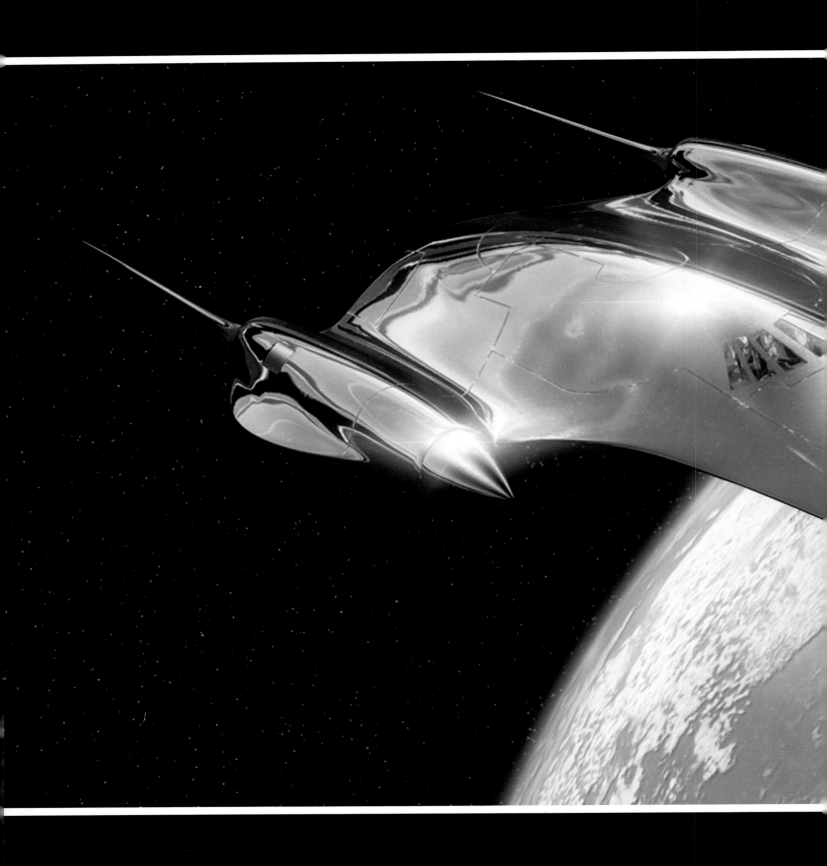

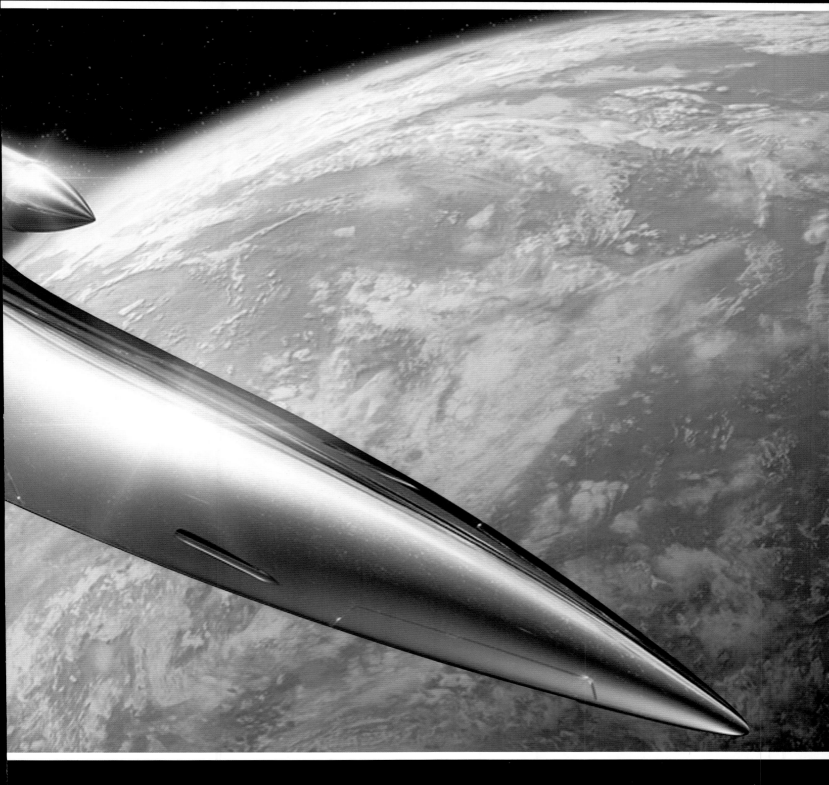

Sleekly futuristic in its design, the Naboo cruiser carries passengers in fine style, but lacks weaponry for attack or defense.

A detailed composite photo taken by NASA's Spirit rover of the Mars landscape shows the potential for extra-terrestrial space travel. Improved propulsion and robotic exploration hold the promise of broadening mankind's understanding of our universe.

Does this mean we are doomed to remain forever tied to our solar system? I do not believe it does. We just have to be a bit more creative about what we mean by space travel. For instance, we might need to forego the idea of a round-trip. If a large-enough craft could be designed to provide a self-sustaining internal environment for a significant population over extended periods, then the need to move such a craft at large velocities diminishes. Imagine intergenerational travel, taking perhaps thousands of years, moving silently throughout interstellar space.

For some, this may be harder to conceive of than building a warp drive. However, I would argue that throughout human history intrepid explorers, and desperate pilgrims, have embarked on what were destined to be one-way voyages to explore brave new worlds. Is it impossible to imagine a situation in the future where life on Earth had become so difficult that even a long one-way voyage to the unknown might be preferable to staying put?

Even if the answer to this question remains a resounding "no," we still have recourse to direct exploration of the universe beyond the confines of our solar system, anticipated to some extent by *Star Wars*. I refer to the charming droids that played a central role in the *Star Wars* galaxy.

If the enemies of space travel are mass and radiation exposure, then one clear solution presents itself. Don't send heavy people; send smaller, intelligent robots. We've seen how the Mars Rovers and the Huygens probe have allowed us unprecedented access to Mars and even a remarkable opportunity to listen to the winds on a distant moon of Saturn. As computer technology continues to improve, along with improved opportunities for miniaturization, I have little doubt that computers will become the astronauts of the future. Their smaller size and diminished demands for maintenance power, combined with the fact that they don't need to be brought back to Earth, make them the clear future choice to boldly go where no machine has gone before. And if, as I expect will happen within the next century, artificial intelligence becomes self aware, then machines will not only be more efficient space travelers, they will be far more capable ones as well.

MAY THE FORCE BE WITH YOU?

Traveling throughout our galaxy with impunity may remain the domain of *Star Wars* and other science fiction movies for the foreseeable future, but even if we must stay confined within our own solar system, we are nevertheless partaking in deep-space adventures that rival anything the *Star Wars* writers have come up with. *Star Wars* may have its ubiquitous "Force," which can be used for good or evil, but in the past decade we have discovered a new force that not only dominates the current behavior of the visible universe on the largest scales, but will also ultimately determine the future of life within it.

By examining the behavior of distant supernovas we have discovered that recession velocity of distant galaxies, hundreds of millions if not billions of light-years away from us, appears to be increasing with time. If general relativity operates on all scales as we think it does, this phenomenon can be possible only if space itself is endowed with a new kind of energy. For, according to the equations of general relativity, if one puts energy into empty space, devoid of all matter and radiation, then unlike the gravitational effects of the energy associated with these latter quantities, the gravitational force associated with energy in empty space must be repulsive. Such a new repulsive force throughout empty space could be small enough to appear essentially invisible on a human scale, but on the scale of the largest known configurations of matter—clusters of galaxies spanning tens of millions of light-years across—this repulsive force could begin to dominate the attraction of matter, causing observable acceleration.

Two questions immediately come to mind as a result of such a revolutionary discovery. First, how did the energy of empty space get there? And second, how will it evolve to affect the future of the universe? Alas, we currently have no idea what the answers might be to either of these questions. They represent the biggest cosmic mysteries in astronomy and physics, and perhaps in all of science. We should not be discouraged by this confusion, however. In fact, nothing excites scientists more than being utterly confused, because it means we have a great deal more to learn about nature. And, as with *Star Wars*, it is much less fun being part of the adventure if you know in advance how it will all turn out...

Supernova: The death of a star, which collapses and implodes after swelling into a red supergiant.

AIR AND SPACE VESSELS

A GALLERY

NORTHROP GRUMMAN B-2 SPIRIT BOMBER

First publicly displayed in 1988, the B-2 nonetheless represents the face of 21st-century military aircraft. Its unusual flying-wing shape makes it superbly aerodynamic, and—combined with a host of new materials and technologies—renders it extremely difficult to detect. This "stealth factor" has become an integral feature of most of the newest aircraft designs under development in the U.S. and abroad.

NORTHROP HL-10 LIFTING BODY

The HL-10 may not look like much, but in 37 flights it set several NASA records. Part of NASA's long research project to design a reusable vehicle that could climb out of the atmosphere and return to Earth, it reached a top speed of 1,228 mph (Mach 1.86) and a maximum altitude of 90,303 feet, beyond the atmosphere's limit. The HL-10's success with power-off landings convinced NASA engineers that the Space Shuttle did not need air-breathing engines.

GETTING AROUND

SCALED COMPOSITES SPACESHIPONE

SpaceShipOne, designed by Scaled Composites, is the first reusable space-craft built by a nongovernmental group. Winner of the $10 million Ansari X Prize in 2004, it made two flights to the edge of space within two weeks, proving the viability of (relatively) low-cost, reusable spaceplanes. Soon you may be able to take a quick joyride into orbit, float around for awhile in zero gravity, and be home in time for lunch.

NERVA

In the 1950s, the U.S. government initiated project NERVA (Nuclear Engine for Rocket Vehicle Applications) to test the feasibility of nuclear rockets. By the late 1960s, several nuclear engines had been test-fired, and NASA had plans for a manned mission to Mars in the early 1980s. The enormous cost of the project and the receding threat of the Soviet space program led to NERVA's cancellation in 1971.

As manned flights to the moon came to an end in 1972, a group from the British Interplanetary Society met to design Daedalus, an unmanned space-craft that would explore not nearby planets, but nearby stars! As planned, Daedalus would have exploded tiny thermonuclear bombs to propel itself across interstellar space, six light-years to Barnard's Star. Total travel time would have been around 40 years.

CASSINI/HUYGENS LAUNCH

Chemical rockets, like the Titan IV-B/Centaur that lofted the Cassini orbiter and Huygens probe, are still the vehicles of choice for getting into space. While they may seem old-fashioned, unmanned missions like Cassini/Huygens generate impressive results. Cassini will continue to study Saturn and its moons for years, while Huygens allowed the world to listen to the sound of the wind on Saturn's second largest moon, Titan.

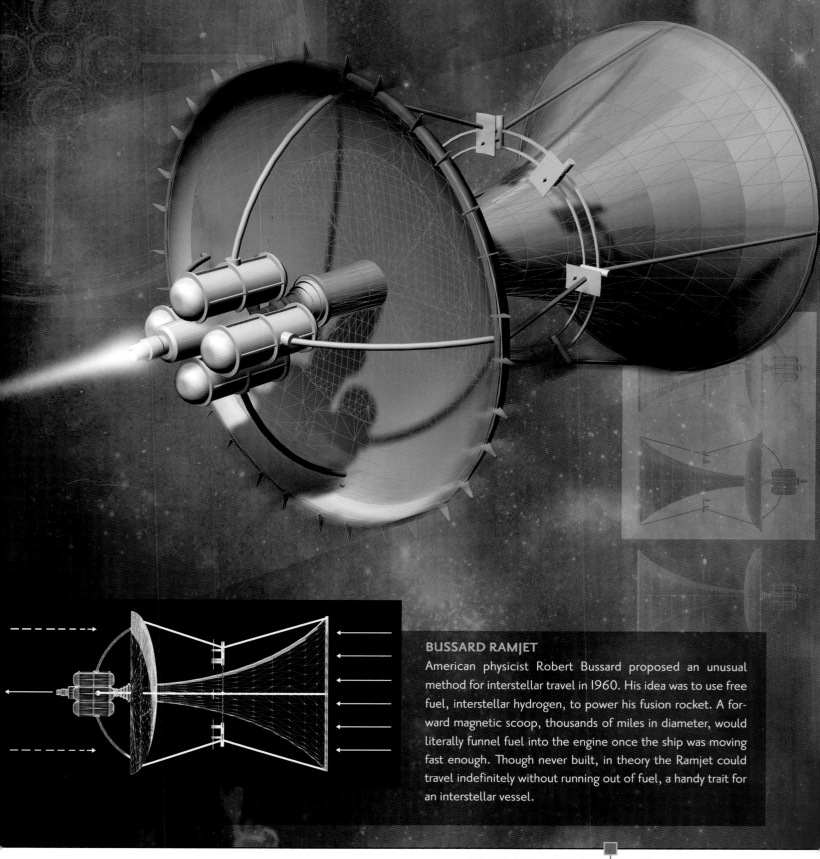

BUSSARD RAMJET

American physicist Robert Bussard proposed an unusual method for interstellar travel in 1960. His idea was to use free fuel, interstellar hydrogen, to power his fusion rocket. A forward magnetic scoop, thousands of miles in diameter, would literally funnel fuel into the engine once the ship was moving fast enough. Though never built, in theory the Ramjet could travel indefinitely without running out of fuel, a handy trait for an interstellar vessel.

MAKING THE JUMP TO LIGHT-SPEED

MARC G. MILLIS

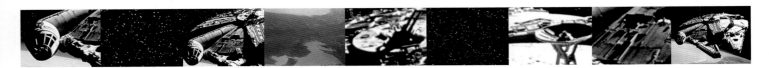

MARC G. MILLIS, Propulsion Physicist and Founding Architect of the Interstellar Flight Foundation. Marc Millis designed ion thrusters and other propulsion systems for NASA; he recently co-founded the Interstellar Flight Foundation to work toward real star travel.

I LIKE THE WAY THAT SCIENCE FICTION TEASES MY BRAIN, PULLING MY THOUGHTS INTO situations I wouldn't otherwise encounter. Provoked by some fantastic device or effect, like Luke Skywalker's levitating landspeeder in *Star Wars*, my imagination begins to brew, contemplating how to actually build such a thing. I enjoy this thought process most—and find it most productive—when it involves a cyclic interplay of imagination, reasoning, and comparison, with knowledge.

I began such playful pondering when I was very young, watching television shows like *The Outer Limits*, *The Twilight Zone*, and eventually, *Star Trek. Star Wars* hit when I was already in college and on my way to studying physics and engineering. All those futuristic devices and twists of perspective teased open my curiosity. As a youngster, I readily accepted whatever premise they provided, but as I grew older I began to notice the occasional logical inconsistencies. Because these required reasoning to sort out, like solving a puzzle, the inconsistencies actually fueled my interest. Toying with these ideas was so engrossing that I was inspired to learn about real physics and engineering to fill in the gaps, and this eventually led me to a career dedicated to shifting the line from fiction to reality.

The inspirational qualities of science fiction should come as no surprise. We know that early rocket scientists such as Robert Goddard and Wernher von Braun were inspired by the science fiction of their day. The allure of rocket travel and its implications for a better world spurred these pioneers to figure out how to make rocketry real. Many of my coworkers and I were similarly influenced by the fiction of our youth—and by real adventures, such as the Apollo program, that grew out of the imaginative musings of our predecessors. There is also, however, a common misperception that science fiction reliably predicts future technologies. Much in fiction hasn't happened yet, didn't happen as envisioned, or will likely never happen. Consider the events of 1969, when men first walked on the moon. Yes, the Apollo crew size and the launch location matched Jules Verne's fiction, but the mission used rockets instead of the author's giant cannon, and, most importantly, people around the world were watching on television. On this day, when the moon was finally won, the Earth was one. The visceral power of that real moment, when Earth's inhabitants shared the experience as one people, was *truly* profound.

Robert Goddard stands next to his invention, the first liquid-propellant rocket, which made history by flying 41 feet in 1926; Goddard became known as the father of modern rocketry.

Opposite: Wernher von Braun, director of NASA's Marshall Space Flight Center from 1960 to 1970, began his career developing the V-2 rockets that Hitler used to bomb London.

The most contentious aspect of science fiction's predictive power arises when it comes to the impossible. Some things, regardless of how plausible they seem in fiction, *might* just remain impossible—sad, but true. Although history sparkles with infamous quotes from scholars declaring the impossibility of such things as nuclear power, heavier-than-air flight, and spaceflight, we tend to forget all the other "impossible" ideas that really didn't work. In retrospect, it's easy to see which is which. Predicting the future, however, is much, much harder. Some predictions can be made by extending the known laws of physics, but the operative word here is *known*. Physics is not dormant. We can reasonably expect further advances, and with each newly discovered law of physics, we'll have new opportunities. But we can't predict what these will be, since accurately predicting what we don't even know we don't know would require the power of clairvoyance.

One of the more passionately debated impossibilities is faster-than-light (FTL) travel—a necessary ability for traveling astronomical distances in ordinary time. In *Star Wars,* it's done with hyperspace. In reality, a wealth of reliable knowledge suggests that FTL travel will remain forever impossible, and, yes, some learned scholars do indeed proclaim it will never happen. (Perhaps they are clairvoyant!) Meanwhile, papers emerge in the scientific literature about wormholes and warp drives, where techniques from Einstein's theory of general relativity theoretically circumvent the light-speed limit. Each paper inspires counter papers and counters to the counters. It is the scientific process in action—exposing and scrutinizing ideas until we discover how Mother Nature truly works.

This cycle of scientific debate is only part of the process of discovery. Before ideas can be subjected to rigorous scrutiny, there have to be ideas to scrutinize. Science fiction can help with this pre-science stage by providing a venue where our thoughts can break from the familiar and toy with the impossible. Arthur C. Clarke, a famous science fiction writer and visionary technologist, put it this way:

CLARKE'S 1ST LAW: When a distinguished but elderly scientist states that something is possible, he is almost certainly right. When he states that something is impossible, he is very probably wrong.

CLARKE'S 2ND LAW: The only way of discovering the limits of the possible is to venture a little way past them into the impossible.

CLARKE'S 3RD LAW: Any sufficiently advanced technology is indistinguishable from magic.

In practical terms, science fiction can provide a context for recognizing new product opportunities, and a setting for "thought experiments." A normal tool of reasoned discovery, thought experiments allow us to explore the consequences of a hypothetical situation in order to see if it makes any sense. If a hypothesis passes the thought experiment, it can be further refined and tested with real experiments. Some thought experiments only progress far enough to reveal logical inconsistencies, a paradox, or gaps in our knowledge,

Hyperspace: A fictional space in which laws of physics may be circumvented allowing faster-than-light-speed travel. In physics, however, hyperspace is a theoretical entity. The theory consists of the idea that our own universe is connected to other universes through wormholes, and all of the universes are found within hyperspace.

but even that represents progress. So, invoking Clarke's 2nd law, and viewing the fictional world with a playful combination of imagination and intellect, we can use visions from *Star Wars* to toy with the impossible and supplement the process of discovery. Doing so takes advantage of three aspects of the film: the story line, which helps us stay focused on the goal; the vividly imagined worlds that help us break from the familiar; and specific images that provide backgrounds for thought experiments. Using some of the most intriguing transportation devices from the film—in particular, Luke Skywalker's landspeeder, Han Solo's *Millennium Falcon*, and the myriad small spaceships that dart around the big screen—we can delve into topics in physics that include basic propulsion, gravity control, and faster-than-light travel.

Ah, the allure of that landspeeder! Gone are the troubles with potholes and hydroplaning. But isn't it curious—*thought provoking*—that Luke's levitating landspeeder only hovers at a low altitude. It doesn't seem to matter how much the craft is carrying, either; it levitates the same whether it's empty or carrying a full load of *Star Wars* characters. Is its levitation height adjustable? The cliff scene in Episode IV presents us with an extreme case to ponder. What would happen if Luke drove that landspeeder over the cliff? Would it maintain its level as the ground dropped away, or perhaps gracefully descend and settle into a new hovering level just above the lower terrain? Or would it plummet out of control, bringing a quick end to our story? Of course, the answer depends on how the speeder works—and that depends on whether we are talking about the movie or *real* levitation.

Herein lies a point of divergence. In the movie, the speeder will perform however the story line requires, and thus create occasional logical inconsistencies. If you want only the fun of the movie, you can ignore these inconsistencies. But in our thought experiment,

Hovering effortlessly in park, Luke's landspeeder uses gravitational repulsion to levitate with no visible means of propulsion or support. The ability to counter the force of gravity continues to elude physicists, remaining the domain of fantasy and science fiction.

All three vehicles, right, have the ability to hover. The blimp at bottom uses lighter-than-air gasses to stay aloft, where it is also at the mercy of moving air currents, which can buffet it about. A helicopter's rotating blades direct thrust to create lift and provide more stable forward movement, though they also create distracting downdrafts. Luke's landspeeder, top, has a clear advantage: It's powered by the force of imagination.

we want to investigate the story's vision of practical levitated motion, a process of exploration that mixes together imagination, knowledge, and reasoning.

We can start by looking at a logical inconsistency. Consider the height at which the speeder levitates. We can infer, based on how other devices performed later in the movie, that the speeder has the ability to adjust its height to rise above obstacles—a desirable feature. But if so, Luke could have raised his altitude to give himself a better vantage point while looking for R2-D2. He didn't. Another inconsistency involves the speeder's jet engines. In one scene the speeder is parked in a strong wind. Somehow, without any anchor or jet thrusting, it keeps itself in place. This implies that the speeder can generate strong sideways forces without the jet engines. Such a feature would be highly desirable, as it could both propel and steer the speeder, making the jet engines superfluous. Of course, this anchoring feature is really a consequence of the special effects hardware, and the advent of digital technology would soon allow George Lucas to create airspeeders that both levitate and dip precipitously. Still, that's not the point here. The point is to let the movie's images, right or wrong, stimulate thought and scrutiny. Such inconsistencies often bug the heck out of me when I'm watching them, but they can also unleash my imagination.

But back to our cliffhanger, where we're still wondering what might happen if the speeder drove over the cliff. For starters, we can compare this situation with a variety of *real* devices that hover: blimps, helicopters, hovercraft, rockets, and magnetic levitation (maglev) trains.

A blimp or hot air balloon uses a huge volume of lighter-than-air stuff—hydrogen, helium, or hot air—to counteract its heavier-than-air parts. It hovers by having the same density (weight-per-volume) as the surrounding air, making it buoyant. If it gets a little lighter, it rises; heavier, it descends. Its levitation is also affected by the surrounding air pressure, but not by the ground underneath. A blimp drifting over a cliff would not change its altitude. This is a desirable feature. But also consider this: Having the same weight as the surrounding air makes the blimp susceptible to being blown around. This is not desirable.

Next, consider a helicopter. By continually pushing down the surrounding air, a helicopter keeps itself up—Newton's third law in action. Since it is using the air instead of the ground, it would have no trouble maintaining the same altitude as it flew over the cliff. The bad news is that it creates a tumultuous swirling of dust, and requires rather large rotating blades to produce its lift. Hovercraft also use downward airflow to keep themselves up, but they take advantage of ground effect so that less airflow is required. The downside, so to speak, is that if the ground were to fall away, our hovercraft would plummet out of control: definitely not a desirable feature. Also, a hovercraft creates an undesirable swirling of dust, though not as tumultuously as a helicopter.

A rocket could maintain its altitude regardless of the terrain underneath, but not for very long. Like a helicopter, a rocket pushes matter downward to keep itself up. Unlike a helicopter, it does not use the mass around it—air—but a limited supply of mass within it, called propellant. That's why rockets can operate in the vacuum of space where there is no air. But a rocket would run out of propellant in just a few minutes. Our landspeeder needs to levitate for at least a few hours…and the fictional speeder even levitates when it's turned off!

Finally, in the case of those maglev trains—well, of course, they are *trains*, and that means they require a track. With our speeder, we want the ability to go anywhere, track or no track.

For a successful outcome, then, we need to make the speeder buoyant, or engineer it to continually push down on some type of reaction mass to keep itself up, or have it create some kind of repulsion against any type of ground. In the first instance, we can use analogies to buoyancy and repulsion to contemplate *gravitational* levitation—toying with the impossible. Presently, the idea of modifying gravity for levitation is beyond known physics, so from here on we are venturing into sheer conjecture.

As it happens, there is another real and curious levitation device touted on the Internet. It goes by the name "lifter," but has many aliases. Contrary to the antigravity claims of many of its proponents, the device pushes down on the surrounding air to keep itself up, much like a helicopter. Instead of using rotating blades, however, it has stationary, high-voltage electrodes that create ion currents, which in turn induce sufficient airflow to lift the device. Because these devices are quite light, they don't require much airflow to stay up. They make for curious experimental gadgets for studying high-voltage ion drift, but would not be suitable for our landspeeder.

Imagine if we could make our speeder buoyant to gravity like a blimp is buoyant to air. What exactly does this mean and what would happen? One perspective involves somehow eliminating the gravitational mass of the speeder, the characteristic that makes it fall in a gravitational field. Mathematically, this can be further explored in terms of energy-per-mass relative to a gravitational potential, but this evokes a curious issue about the speeder's inertia. So far, all observed physics indicates that the *gravitational* properties of mass (that which makes mass fall) and the *inertial* properties of mass (that which resists changes in motion) are exactly equivalent. If this "equivalence principle" were true for our fantasy speeder, it would mean that the speeder would also have zero inertia. The slightest force would send it accelerating dramatically away. Although this sounds really cool, it would also make the vehicle extremely hard to step into—definitely not cool.

Ignoring the inertial issue, consider this necessary event of stepping into the vehicle. What happens when we put normal matter—like our hero, Luke—into the speeder? We know that the speeder hovers at the same level, empty or full. But how will the speeder compensate for the added mass? One perspective is that it further reduces its *own* gravitational mass (now less than zero) instead of altering Luke's gravitational mass (a desired feature, at least from Luke's point of view). It's fair to assume that this process requires some energy expenditure. Visually, the scene in which Luke hops out of the speeder, causing it to bounce up and down a little, applies to this situation. Although the bouncing is really a consequence of the long counterweighted arm on which the landspeeder was mounted, it is nonetheless thought provoking, and introduces the mathematics of oscillations as another analysis tool. Just for fun, imagine that the speeder's gravity adjustment got stuck at the setting for carrying Luke's mass just as Luke stepped out. Would the speeder go shooting skyward, having too much gravitational negation?

These are only some of the opening questions that lead to deeper inquiries, but rather than launching into more detail, let's further confuse the matter with yet another levitation method: What if the bottom of the speeder had the ability to block the force of gravity? Would this mean that all objects above the speeder would become weightless? Imagine parking this speeder under just the edge of a Ferris wheel, so that only the section of the Ferris wheel above the speeder would weigh less. The whole wheel, with one side now lighter than the other, would begin to rotate. This requires energy, which we presume must come from a cause—our speeder's gravity-blocking device. Just as with gravitational buoyancy, this energy-transfer situation provides a mathematical opportunity for more rigorous and logical scrutiny.

> Just for fun, imagine that the speeder's gravity adjustment got stuck at the setting for carrying Luke's mass just as Luke stepped out. Would the speeder go shooting skyward, having too much gravitational negation?

What about repelling gravity? In the *Star Wars* world, the speeder is described as using repulsorlift technology. I have no idea what this means, but then, that really doesn't matter, since we're making stuff up to see what happens. Building on the analogy of magnetic repulsion used for maglev trains, consider hypothetical gravitational repulsion. The conceptual shift from magnetism to gravity ensures that our speeder can work over any terrain, not just over special magnetic track.

Immediately, however, this introduces a challenge. Gravity and magnetism, despite some similarities, are quite different. Magnetic forces, which are side effects of moving electrical forces, are characterized by the following facts: First, electrical charges come in two polarities, plus and minus. This gives us the choice of creating net forces from unequal portions of the two polarities, or zero force using equal portions, meaning we can turn electrical forces on and off. Second, like charges repel, while unlike charges attract, which gives us the choice of creating repulsive or attractive forces. And third, some materials conduct electricity, while other materials block it, allowing us to construct devices to channel electricity

Equivalence Principle: Einstein's principle of equivalence states that the (local) effects of a gravitational field are identical in all respects to the effect of uniform acceleration. It is a central principle in the theory of general relativity.

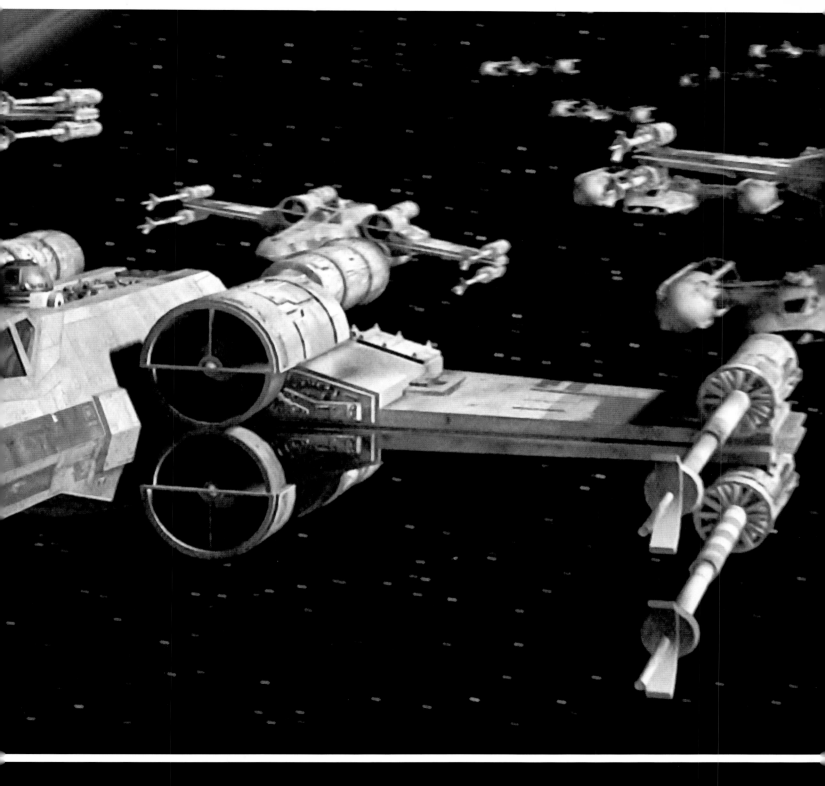

In a seemingly futile mission to destroy the Empire's Death Star, Rebel forces head out in a fleet of X-wing fighters, with wings closed in cruise position.

GETTING AROUND

Paul Moller stands in front of three of his flights of fancy, all of which utilize vertical takeoff and landing technology (VTOL). Moller sees a future in which these low-maintenance, versatile vehicles may replace the car; the M400 Skycar (center) is in its final testing stages, with FAA certification expected by early 2007.

exactly where we choose. Gravity, on the other hand, has only been found to attract, and—if that wasn't disappointing enough—no such things as insulators and conductors of gravity exist. Gravity appears to permeate everything with no way of being shut off or redirected. The only known exception is from accelerating ultra-dense matter at relativistic speeds, where magnetic-like gravitational forces come into play, but even here the results are meager and the energy expenditures enormous.

We also know that gravity and electromagnetism are coupled phenomena, meaning that one affects the other. Our proficiency at manipulating electromagnetism hints that we might be able to use this mastery to affect gravity, and in principle this is true. However, our *present* knowledge does not give us a practical way to negate, block, or repulse the pull of gravity.

In terms of Einstein's general relativity theory, gravitational effects can be created by the presence of energy, even electromagnetic energy. The bad news is that Einstein's theory is cast in terms of defining geodesics, the natural paths that all free-falling objects will follow, not just our little landspeeder. By analogy, let's say we want to go from point A to point

B. The general relativity approach would have us supply enough energy to reshape the entire space-time terrain to create a downhill effect where we (and everything in the vicinity) would naturally fall from A to B. This, obviously, is overkill. We want something more akin to a jeep that can propel itself across the natural terrain of space-time without having to launch a major civil engineering project to build new space-time roadways.

And finally, we could consider the mechanical analogy of simply raising the speeder with a jack, which only requires the energy of a single person. But this solution assumes a firm mechanical connection to the ground, which is not true levitation. Magnetic levitation and acoustic, or sound-wave, levitation have been achieved, but they only work within the confines of an external supporting device, instead of being produced by the object we want to levitate. And that's about it—presently no known way exists to create the repulsion effect of Luke's landspeeder.

So where do we go from here? If we aim to create a vehicle that can levitate without wheels, rockets, or downward thrusting of air, we need some new tricks regarding the natural forces of gravity, or perhaps just a novel approach. Here are some leads and unknowns to ponder: the equivalence principle; general relativity and basic mechanics as they pertain to gravity; the energy and power required to lift an object; the energy and power required to shield half a Ferris wheel; and the notion of gravitational conductors and insulators. Even when possibilities turn out to be dead ends, the learning process itself remains rewarding.

Normal gravity occurs so commonly onboard science fiction spacecraft that I'll bet most of the audience never realizes just how profoundly amazing this would be. Moving in space, whether in orbit or coasting across its vast distances, a spacecraft and all its innards are in free-fall. Crew members float inside their cabin, weightless. In the filming of science fiction movies, however, this zero-gravity effect is very hard, and expensive, to simulate. Instead, virtually all fictional spacecraft seem to synthesize normal Earth gravity within them, conveniently matching normal studio conditions.

Consider the real effects of synthetic gravity: Longer space missions would become possible, as this would alleviate the detrimental health issues caused by prolonged exposure to microgravity, such as bone-mass loss. And if we can create synthetic gravity, we can also turn it on and off, meaning we could adjust it to different levels of intensity. And why not orient it sideways instead of just up and down, again with varying intensity? Gravity could then be used to move objects around in multiple directions, not just to hold astronauts against the cabin floor.

Next, extend the idea's scale to the very small and the very large. Synthetic, localized gravitation could be used to pump water, air, or a whole host of things. If its intensity could be quickly oscillated, it could be used to vibrate objects without contact—perhaps even in ways tailored to the manufacturing process. If the effect could be applied with different intensities in different places and at different angles, you could even use it to shape and perhaps cut materials. The manufacturing and material-handling possibilities are enormous.

Geodesics: the shortest line between two points on a mathematically defined surface (as a straight line on a plane or an arc of a great circle on a sphere).

For recreation on Earth, you could turn the effect upside-down to create zero-gravity rec rooms. Consider the franchise possibilities!

But wait, there's more! All these variations assume that synthetic gravity is produced inside a box—between floors and ceilings, for instance. What if the effect could be turned inside out? Rather than accelerating objects *inside* the box, why not use synthetic gravity to accelerate objects *outside* the box? This ability, depending on its effective range, could push spacecraft away from, or toward, nearby planets. It could even accelerate a spacecraft in any direction, anywhere, given a wide-enough effective range. Such an external gravity generator would become the engine for the spacecraft. Unlike a rocket, it would not need propellant, but would rely instead on the surrounding matter of the universe to push against. Of course, the spacecraft would still need an energy supply to power all maneuvers, and here the mathematics of power, energy, and motion would come into play. Aside from the fact that there is no known way to do this, the notion of interacting with universal mass over astronomical distances is not a trivial one, and related concepts such as Mach's Principle, about the connection between matter and inertial frames, are still debated today. The notion of "space drives" has even been introduced in professional journals, albeit at very early stages of exploration.

When science fiction portrays synthetic gravity onboard its spacecraft, it's reasonable to expect that this same technology could propel its vehicles, rendering jet engines superfluous. Once again, however, we are presented with a logical inconsistency, since many of the *Star Wars* vehicles apparently use jet engines with glowing exhausts—even in the vacuum of space!

One exception in *Star Wars* to those glowing thrusters is the TIE fighter, where TIE stands for "Twin Ion Engines." In the film, these engines are located in the rear circular part of the ship and glow with a red light. In reality, there is indeed a type of electric rocket called an "Ion Engine." Rather than combusting chemicals that spew dramatically from a nozzle, like the space shuttle's rockets, ion engines use high voltage to accelerate a charged gas (plasma) out of a modest round screen. Although an ion thruster produces dramatically less force than its better-known chemical cousin, its propellant efficiency is reciprocally much greater, meaning it can travel greater distances on a single tank. And for journeys farther than Mars or thereabouts, the ion engine can even get there faster, thanks to its continuous thrust.

In addition to ion engines and flaming chemical rockets, concepts exist for other propulsion methods using nuclear energy, as well as fledgling work into antimatter. More than science fiction, antimatter has been around for decades as a by-product of high-energy physics experiments. Technology has reached the point where portable anti-proton traps are being explored in the lab, while analysts contemplate how to use antimatter in the very limited quantities that can be produced today. All of these concepts, however, are still based on the principle of a rocket, which is limited by how much propellant it can bring along for the

Mach's Principle: Assertion that the inertial effects of mass are not innate in a body, but arise from its relation to the totality of all other masses, i.e. to the universe as a whole.

Antimatter: Matter consisting of elementary particles that are the antiparticles of those making up normal matter.

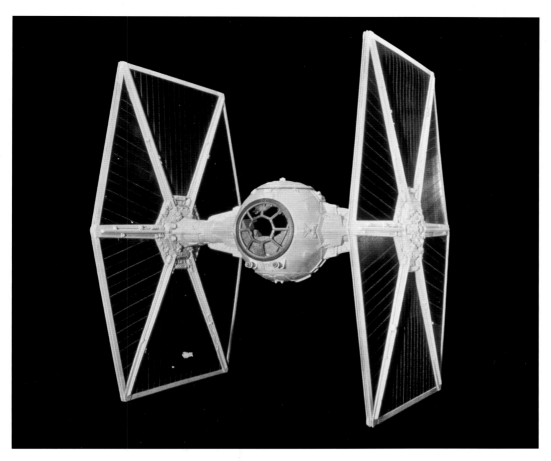

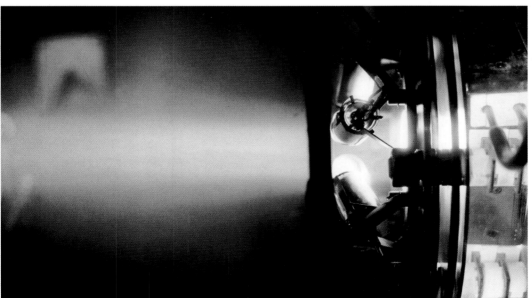

In 1977, George Lucas created the TIE (Twin Ion Engine) fighter, top, for the Empire. Bottom: NASA first used its ion engine, developed in the 1950s, on the Deep Space 1 probe launched in 1998.

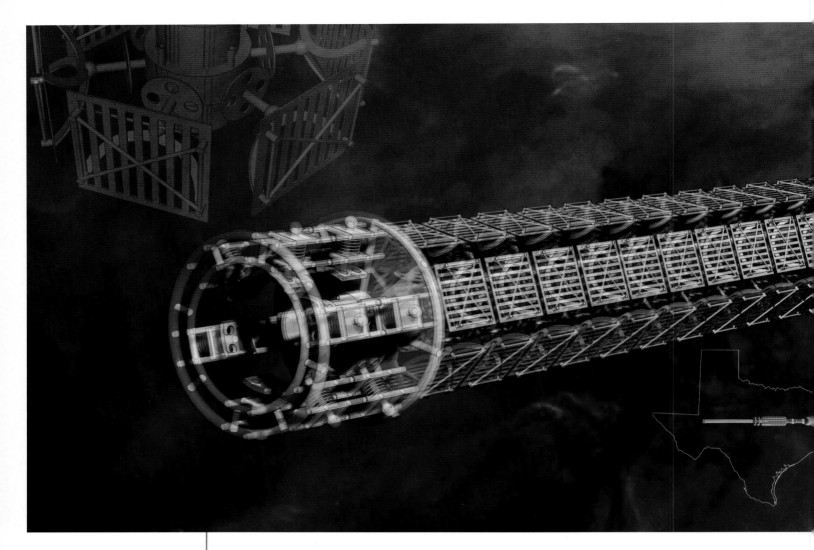

journey. The farther or faster a rocket needs to go, the more propellant it requires, which means that even more propellant will be needed to propel the extra propellant. It all adds up, exponentially.

When it comes to interstellar flight, rockets are ridiculous. The best available rockets would still take tens of thousands of years to reach our nearest neighboring star. We could get there faster using rockets that are still on the drawing board, but even here the cost of a faster trip is more propellant. To bring trip times down to less than a century, we need propellant quantities akin to the size of the moon, or even the sun—most unacceptable! There are other concepts, called light sails, which do not need propellant. Pushed by either a powerful laser (that does not yet exist) or sunlight, sails made of ultra-thin material spread out over several kilometers just might make a faster interstellar probe—and, in fact, Episode II *Attack of the Clones* features a solar sailer, in which Count Dooku makes his escape from the Jedi. But, even here, we're still looking at extremely long flight times or enormous power requirements. Right now, the best hope for practical interstellar travel lies

GETTING AROUND

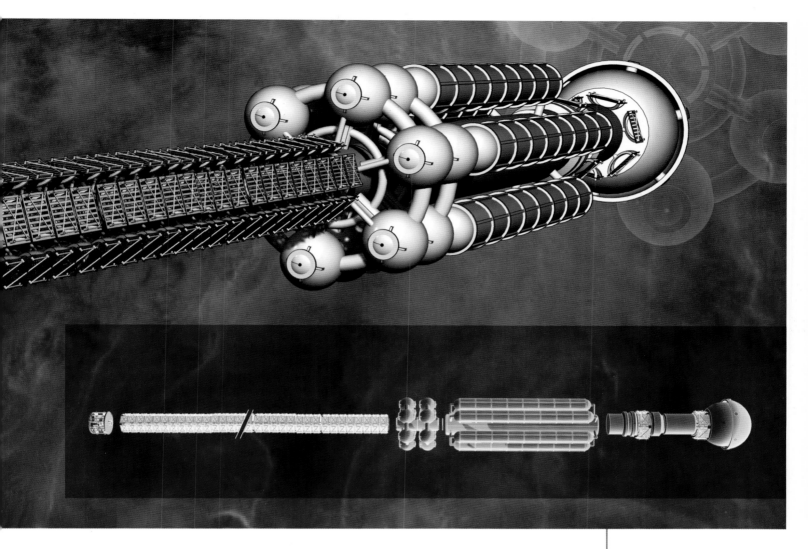

in physics breakthroughs—the kind that might give us space drives similar to those of today's science fiction.

Where does that leave us? All those loose ends about gravity control for the landspeeder apply to deep-space travel as well, in addition to such issues as conservation of momentum, the challenge to produce net force, and the need for asymmetric fields. Even if a gravity space-drive was discovered, we'd still have the problem of covering astronomical distances in reasonable times. For that, we must turn to the challenge of FTL travel.

Right now, I'm sad to say, faster-than-light travel looks like an impossibility, whether we're talking about brute-force acceleration within space-time, or modifying space-time to circumvent the light-speed limit. The lessons of special relativity rule out the brute-force approach: Even though applying more energy proportionally increases speed, this only works at low speeds; it gets harder and harder the closer we get to light-speed. In fact, the energy required to speed matter up to light-speed is infinite. The other approach uses general relativity to modify space-time, but all its variants seem to lead to the paradoxes of

Antimatter mixed with matter could be the most efficient rocket fuel possible. A spacecraft with a matter/antimatter drive could reach speeds approaching two-thirds the speed of light.

time-travel: a real showstopper. A third approach from quantum physics involves "entanglement" and "non-locality," but these are subject to different interpretations, so much so that it is way too early to determine their applicability.

So what possibilities does fiction offer us? In his 1928 book, *The Skylark of Space*, E.E. ("Doc") Smith may have been the first author to mention the concept of faster-than-light travel. Interestingly, this came two years after Robert Goddard made his first successful liquid rocket experiment. Others credit the idea to the science fiction of John Campbell; in his 1931 book *Islands in Space,* Campbell used the term "hyperspace" for his version of FTL travel.

With the concept of hyperspace, the approach used in *Star Wars,* the spacecraft somehow enters a different realm—the hyperspace—where different speed limits apply. Once the vehicle has traversed the appropriate distance, it reemerges into normal space-time. Through hyperspace, the craft can cover more distance in less time. Unlike warp drives and wormholes, hyperspace has not yet been explicitly discussed in legitimate scientific journals.

Warp drive, one of the most predominate concepts for FTL travel, circumvents the limit of how fast an object can move through space-time by moving space-time itself. It somehow creates a moving chunk of space-time that then carries a stationary vehicle along inside it. Different rules govern the motion of space-time than those governing how objects move within space-time. Presently, this concept must still be considered science fiction, although healthy debate has been emerging in legitimate scientific journals. A 1951 letter from Malcolm Gibbs in *Marvel Science Stories* may contain the first appearance of the term "warp drive," and before that, the closely related term "space warp" appeared in 1935 in Nat Schachner's *The Son of Redmask*. In scientific journals, warp drive first appeared in 1994, in a letter published in *Classical and Quantum Gravity*.

The Jedi answer to faster-than-light travel lies in a special transport booster, called a Syliure-3 long-range hyperdrive module. Above, Obi-Wan Kenobi docks his starfighter into a waiting transport ring, exponentially increasing his personal vessel's capabilities.

We can toy with this notion of hyperspace by bringing in an analogy to a real speed-breaking event: the 1947 shattering of the sound barrier. Here's an important point: Sound did not break the sound barrier! In fact, the sound barrier succumbed to a rocket-plane made of atoms and molecules held together with electromagnetic forces. Electromagnetism is the same stuff as light, so, in a sense, light broke the sound barrier. When it comes to breaking the light-speed barrier, we are essentially trying to exceed the speed of the very property that connects our matter together—a very different problem.

Electromagnetism governs the connections of atoms and molecules, whose mechanical vibration produces sound. The vibrations of sound are basically a slower mechanical side effect of the much faster electromagnetic forces that hold us together. How much faster is light than sound? About 150 thousand times faster. Imaging trying to move a piece of sound, say a conversation, faster than the speed of sound. Actually, this happens all the time. Converting the sound to light, such as radio waves, allows it to travel faster and farther, and it can then be converted back into sound at the final destination. Radio and television do this constantly. In another approach, imagine putting the source of the sound in a box that can move faster than sound—a supersonic aircraft, for example.

Now to delve into wild conjecture: Apply this analogy to transporting ourselves by first entertaining the idea that our electromagnetic existence is just a slower side effect to some other underlying faster phenomenon, called hyperstuff (for lack of a better name). In this construct, light is a slower side effect to our unknown hyperstuff, in a similar fashion to the way sound is a slower side effect to light. Moving even deeper into this science wilderness, consider a hyperstuff vehicle that can carry our electromagnetic essence through hyperspace, akin to the way a supersonic aircraft can carry sound. Although this conjecture requires quite a stretch, two thoughts led me to it: the analogy of breaking the sound barrier with matter, and the question, provoked by scenes in *Star Wars,* of what hyperspace might actually *look* like.

The notion of leaving our own physical space to enter hyperspace gives rise to some interesting questions. What effects would happen in our own space at the moment that our spaceship disappeared into hyperspace? What phenomena might remain of the spaceship, if any, and how long would they take to fade away? Similarly, when the spaceship suddenly reappeared, how long would it take a distant observer to notice its image and its gravitational field?

Such notions of hyperspace, faster-than-light travel, gravity control in spacecraft, and the design of landspeeders remain primarily within the realm of science fiction. Some related work is just beginning to enter professional discussions, sometimes without the giggles. For a few years, NASA conducted related research, and I published those findings as a technical paper in 2004. On the more near-term possibilities of interstellar flight, a wealth of factual, scientific information exists. And, of course, we still have science fiction—the stuff of inspiration, and dreams.

Electromagnetism: Relates to the magnetic field generated around a conductor when current is passed through it; magnetism produced by electric charge in motion.

MAGLEV TRAINS

SAM GUROL AND HIROSHI NAKASHIMA

SAM GUROL, Director of Maglev Systems, General Atomics. At the Electromagnetic Systems Division of General Atomics, Dr. Gurol leads a team working on the development of an urban maglev system.

HIROSHI NAKASHIMA, Deputy Director General, Maglev System Development Division, Central Japan Railway Company. One of the early proponents of maglev technology, Dr. Nakashima has devoted his career to achieving maximum-speed rail travel without sacrificing safety.

superconductivity: The flow of electric current without resistance in certain metals, alloys, and ceramics at temperatures near absolute zero, and in some cases at temperatures hundreds of degrees above absolute zero.

In 1993, a visionary group of Pennsylvania business people concluded that an elevated magnetic levitation system could offer a cost-effective, long-term solution to Pittsburgh's congestion problems, and might even bring about a paradigm shift—a revolution in urban transportation. Maglev transportation can handle steep grades, it's quiet, and, as an elevated system, it avoids the expense of tunneling. Looking to the future, Pittsburgh's General Atomics (GA) Urban Maglev project put together a team with expertise ranging from the first stages of planning to the last steps of implementation.

In 2000, the Federal Transit Administration awarded GA funding to develop the basic concept of urban maglev and its applications in the American market. Preliminary efforts focused on selecting the methods of levitation and propulsion. After reviewing state-of-the-art systems, the GA team selected "Inductrack" technology, which makes use of permanent magnets and has a fairly large air gap between vehicle and track, indicating less stringent guideway construction tolerances.

The basic vehicle consists of two chassis units, connected by an articulation device that allows the train to negotiate tight turns. Entirely elevated, the system operates automatically, without a driver. A linear synchronous motor (LSM) mounted on the track provides propulsion to the vehicle, which can be configured into various desired lengths. LSM propulsion ensures energy efficiency, as it powers only those sections of track where the vehicle is located.

Levitation is achieved through arrays of permanent magnets beneath the vehicle. When the train is in motion, their magnetic field generates so-called eddy currents, which move in a direction that interacts with their applied magnetic field, producing forces that levitate the vehicle. Below, the electrically conducting track resembles a ladder with closely packed rungs. The train's permanent magnets are configured in a Halbach array: magnet cubes measuring about 5 centimeters per side arranged in a linear fashion along the length of the vehicle, with their polarity changing by 45 degrees from one magnet to the next. This configuration results in a sinusoidal magnetic field focused on the track. The train starts off on polyurethane wheels; lift force increases with speed, until the vehicle levitates at about 10 km/h.

The track-mounted LSM generates a moving magnetic field whose speed is determined by the frequency of the applied current. Imagine ocean waves carrying a surfer to shore: The LSM's magnetic wave locks onto the magnets on the vehicle, carrying it along in a synchronous fashion. In this passive system, levitation results from the forces generated by the vehicle's forward motion. Should propulsion power be lost, the vehicle simply coasts to a landing.

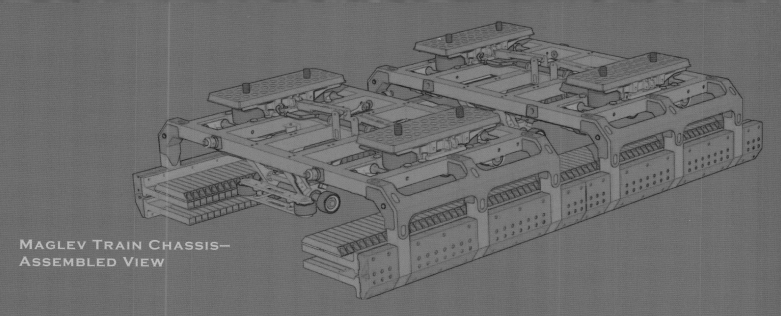

Maglev Train Chassis— Assembled View

In March 2003, General Atomics broke ground on a test facility with a 120-meter-long, full-scale track featuring both straight and curved sections. Three months later, a full-scale test vehicle arrived, with a chassis composed of upper and lower Halbach arrays, additional Halbach arrays for the propulsion system, auxiliary wheels, and secondary suspension components, as well as water tanks for varying the weight load. Levitation, propulsion, and control systems have all been tested, and in 2005 GA will test the capabilities of the entire system at speeds of up to 10 meters per second. Our goal is to validate ride quality, takeoff and landing characteristics, and projected efficiencies, among other features.

Finally, we hope to build a demonstration system at California University of Pennsylvania, about 60 miles southwest of Pittsburgh, with 7.4 kilometers of track and three stations; this system will connect the university and the Borough of California. The challenges of a seven-percent grade and the weather conditions of western Pennsylvania should enable us to demonstrate the climbing and ride-quality capabilities of a maglev system.

We see maglev not only as the ideal solution to traffic congestion, but also as an answer to the problems of global warming and dwindling natural resources. It's quiet, fast, and environmentally friendly; it can negotiate steep grades and tight turns; and it provides an elegant vision of the future: an elevated grid of maglev trains, efficiently connecting all areas of an urban landscape, from the city center to the suburbs, local and distant airports, and beyond.
—SG

As deputy director general of maglev development for the Central Japan Railway Company, I've been working for nearly 40 years on developing technology that ultimately will result in trains capable of traveling at speeds of 500 km/h and more. This technology uses superconducting magnetic levitation—maglev—to produce a whole new railway system in which the motive power for the train comes from the tracks. **Superconductivity** keeps the trains afloat, totally suspended above the ground. No contact with the tracks means none of the friction that normally occurs when train wheels run on a steel rail.

Shanghai, on the east-central coast of China, boasts an operational maglev train service. The city is one of the first in the world to adopt this revolutionary train transit technology.

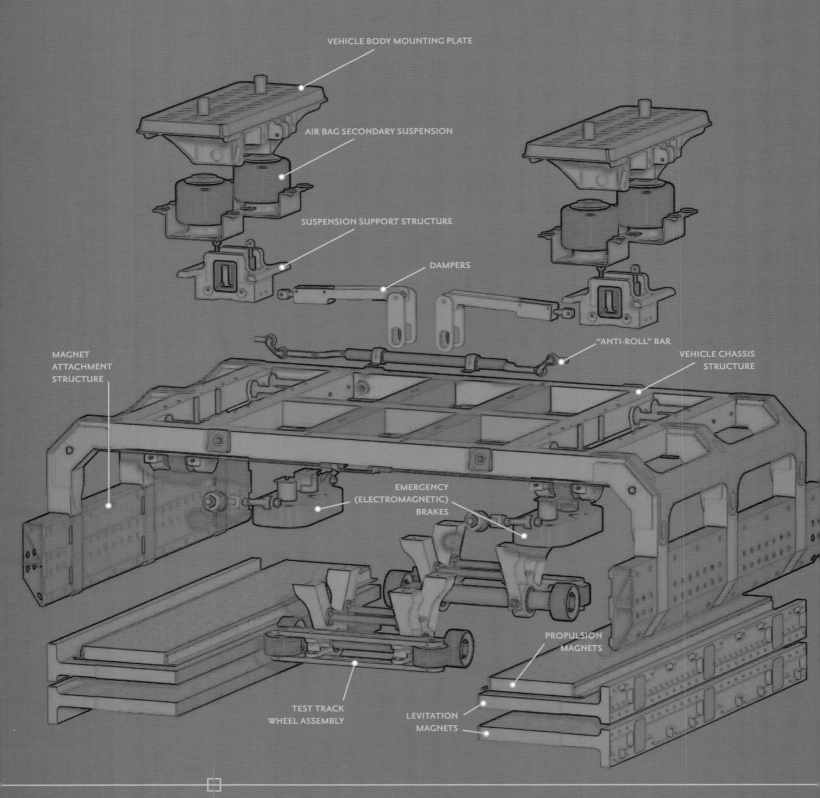

VEHICLE BODY MOUNTING PLATE

AIR BAG SECONDARY SUSPENSION

SUSPENSION SUPPORT STRUCTURE

DAMPERS

MAGNET ATTACHMENT STRUCTURE

"ANTI-ROLL" BAR

VEHICLE CHASSIS STRUCTURE

EMERGENCY (ELECTROMAGNETIC) BRAKES

PROPULSION MAGNETS

TEST TRACK WHEEL ASSEMBLY

LEVITATION MAGNETS

Magnetic levitation has already cleared most of the hurdles that stand between the drawing board and employment in general service. The challenges now involve issues of reliability and operating expenses. Of course, we had to overcome some major engineering challenges to achieve an operational maglev test track; the greatest involved getting the superconducting electromagnets to perform adequately, as the coils had to withstand a very strong electrical current running through them. We had to combat a phenomenon called a "quench" on the magnets which effects their superconductivity, and there were also issues with the magnets' weight and the size of the refrigeration unit—originally larger than the vehicle itself.

By now, we've conducted hundreds of tests without any problems—the technology works, day in and day out. Superconducting magnets on the train respond to ground coils on the guideway; one kind of ground coil affects propulsion, and the other affects levitation. Liquid helium keeps the superconducting coils cool. Three phases of alternating electric currents, from an electric inverter in the substation, are supplied to the propulsion coils. When the vehicle is underway, superconducting magnets induce electric currents in the ground coils for levitation; between these induced currents and the superconducting magnets on board, sufficiently strong magnetic forces levitate the vehicle up to ten centimeters off the ground.

We still need to improve performance, to learn how fast we can run the train while guaranteeing a safe, comfortable ride for passengers and minimizing pollution and noise. The maximum maglev speed, recorded in 2003, is 581 km/h; our MLX-01 test train currently holds the world's record for fastest train, though we believe we can increase this speed. Our goal is to offer maglev train service between Tokyo and Osaka, with links to a central railway. At speeds of 500 km/h, our train would travel between these two cities in an hour, as opposed to the standard two and a half hours.

The average rail speed now between Tokyo and Osaka is 270 km/h, with a top operational speed of about 300. Maglev trains will have an operational speed averaging 500 km/h. The ride is very stable; in many ways, the sensation compares more to flight than to traditional train travel. When the train begins to levitate at 180 km/h and the wheels retract, the motion grows considerably smoother and quieter. I've made about 40 test runs, and I'll never forget when we first achieved our goal of 500 km/h. Of course, we were confident we could do it—but when the speedometer finally clocked up to 500, it was such a relief! I remember thinking, "Oh, we've come such a long way since I started in this field."

Like so much in science and engineering, maglev began with somebody reading about an idea and saying, in effect, I wonder if I could use this to do such-and-such? In 1966, two American scientists, James Powell and Gordon Danby, wrote about a high-speed vehicle levitation system using superconductive magnets. Back then, no one knew much about superconductivity, so it took a leap of faith to research something that sounded like science fiction.

And now, some 40 years later, superconductive magnets are everywhere, in systems ranging from our maglev train to magnetic resonance imaging (MRI), used in hospitals worldwide. I expect to see power plants incorporate superconductive technology in their generation and distribution operations, and the possibilities only go on from there. Someday in the near future, superconductivity will be as common in our daily lives as electricity is today.   —HN

REPUBLIC GUNSHIPS

Look closely at most *Star Wars* vehicles and you can find a real-world inspiration for it. In the case of the Republic gunship, that craft is the military helicopter. Able to drop a squad of soldiers right in the fight, armed to take out targets much larger than itself (but lightly armored and susceptible to ground fire), the Republic gunship has capabilities like the ubiquitous Huey helicopters of the Vietnam era.

STAR WARS
VESSELS
A GALLERY

TRADE FEDERATION MULTI-TROOP TRANSPORTER (MTT)

The MTT can safely deliver an entire battalion of battle droids to wherever they're needed and deposit them in formation, ready for action. The droids are transported in folded position to conserve space. The MTT uses the repulsorlift technology that provides lift for most floating vehicles in the *Star Wars* galaxy; these vehicles can run over you without causing injury, like the MTT that runs over Qui-Gon Jinn and Jar Jar Binks in Episode I.

GETTING AROUND

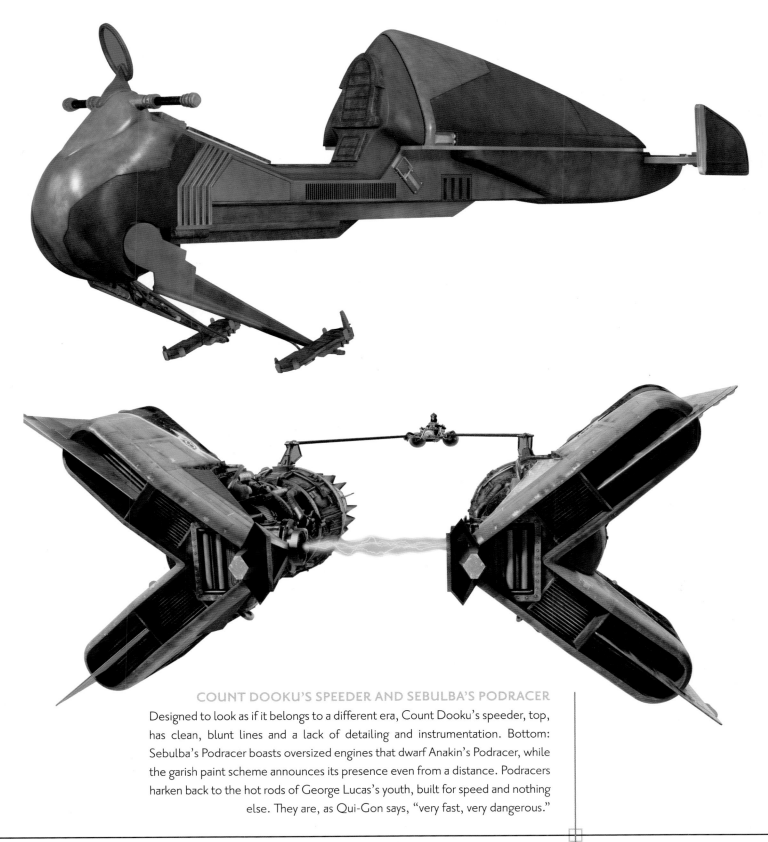

COUNT DOOKU'S SPEEDER AND SEBULBA'S PODRACER

Designed to look as if it belongs to a different era, Count Dooku's speeder, top, has clean, blunt lines and a lack of detailing and instrumentation. Bottom: Sebulba's Podracer boasts oversized engines that dwarf Anakin's Podracer, while the garish paint scheme announces its presence even from a distance. Podracers harken back to the hot rods of George Lucas's youth, built for speed and nothing else. They are, as Qui-Gon says, "very fast, very dangerous."

GETTING AROUND

JEDI STARFIGHTERS AND REPUBLIC ARC-170S

Seeking to bridge the gap between the prequels and the original trilogy, the concept artists of Episode III worked very hard to prefigure the lineage of vehicles that appear in Episodes IV-VI. Nowhere is that more obvious than in the design of the Jedi starfighters and ARC-170s. They combine the flowing lines and graceful curves of the vehicles in Episodes I-II with the more angular, modular look of the ships from Episodes IV-VI.

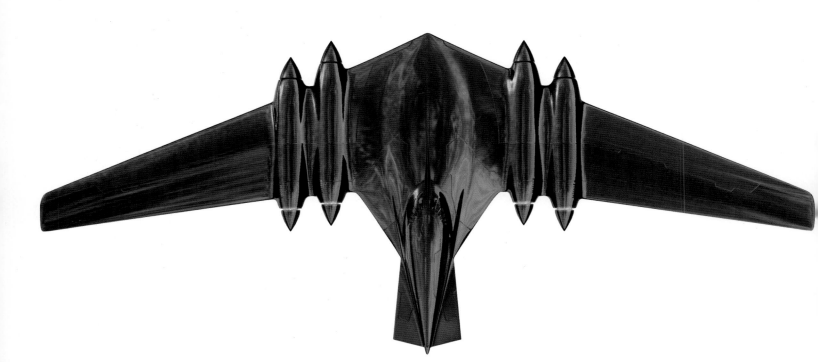

TANTIVE IV AND NABOO CRUISER

The *Tantive IV*, top, was the first spaceship seen in Episode IV and the last to appear in Episode III, where it boasts a number of "beauty panels." By the troubled times of Episode IV, the craft has lost most of its cosmetics and added more guns. Bottom: Throughout Episodes I-III, Naboo's vehicles are graced with unbroken surfaces and heavy chrome to denote royalty. Senator Amidala's cruiser sports chrome in honor of her previous service as queen.

GENERAL GRIEVOUS'S WHEEL BIKE

One of the most unusual hybrid vehicles in the *Star Wars* galaxy, the Wheel Bike designed for General Grievous is capable of using its central spiked wheel for mobility; it can also deploy four jointed limbs to simulate a running gait. Though it lacks repulsorlift capability, it still manages to ascend or descend nearly vertical surfaces, thanks to its sharp spikes and clawed feet.

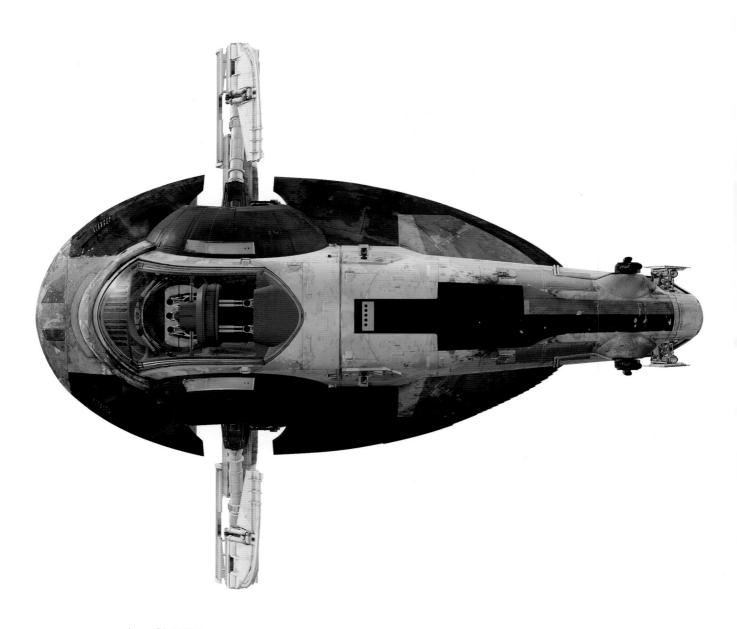

SLAVE I

Like the *Millennium Falcon* and *Tantive IV*, *Slave I* has a name and a long history in the *Star Wars* galaxy. In Episode II, *Slave I* transports Jango Fett and his clone "son" Boba to Geonosis. The ship reappears in Episode V in the hands of the now-grown Boba Fett, who uses it to pursue and capture Han Solo for the bounty from Jabba the Hutt. The owners of *Slave I* take good care of it; aside from the paint scheme, it appears little changed from Episodes II to V. This ship lifts straight up in takeoff, then rotates forward ninety degrees to flight mode.

TRADE FEDERATION GUNSHIP AND ARMORED ASSAULT TANK (AAT)

Geometry plays a role in *Star Wars*: A triangular profile denotes a Republic/
Imperial vessel. Behind opaque windows, bad guys will be lurking, and round
vessels are also usually bad; this Trade Federation gun platform, top, is saucer-
shaped when viewed from above. Bottom: The Trade Federation deployed the
AAT during the Battle of Naboo in Episode I. Its repulsorlift allows it to pause
on a steep incline without sliding down the hill.

VISUAL CONCEPTS IN STAR WARS

ALEX JAEGER

ALEX JAEGER, Industrial Designer and Visual Effects Art Director, Industrial Light & Magic. Alex Jaeger works as a designer and art director for graphic novels, and as a freelance agent within the film industry, as well as for Industrial Light & Magic.

THE DESIGNERS OF SOME OF THE BEST KNOWN FEATURES OF THE *Star Wars* GALAXY— spaceships, Podracers, droids—deliberately make them all look as if they actually work, so that viewers won't require a lot of explanation to believe in what they're seeing. In other words, if the design works, the feature in question will look like it works too, and will be accepted at face value. From the hot-rod feel of Joe Johnston's X-wing and Y-wing starfighters and his *Millennium Falcon*, to the opulent elegance of Doug Chiang's Naboo starfighter and cruiser, *Star Wars* has a rich library of cultures, villains, heroes, and the time span to make it all work.

Landspeeders and spacecraft in the *Star Wars* movies represent an extension of our cars, which in turn stand for personal freedom. But the great thing about the way these craft are portrayed in the films is that they're simply there—part of the landscape, just as our own automobiles exist as ordinary features in our lives. The *Star Wars* speeders don't call attention to themselves; presented as nothing more than necessary modes of transportation, they don't require any special introduction. Of course, some *Star Wars* vehicles can impose a threat through their presence alone. Consider the Star Destroyers: ominous, pointy, lumbering objects that command your attention when they appear behind you. Or the TIE fighters, announcing themselves with a roaring scream as they pass by. X-wing starfighters portray an aerobatic personality with their S-foils closed, and a capable threat when the S-foils move to attack position. It's all part of the design.

Transportation can occasionally become a character in its own right, assuming a larger role with specific "quarks" and foibles. Take the *Millennium Falcon*, for instance. It has a history that it literally wears, and, just as important, it has a name of its own, as opposed to the generic titles of most ships in the films. Like famous real-world craft—the *Memphis Belle* or *Enola Gay*, for instance—the *Millennium Falcon* is special; in fact, it's the only one of its kind. We learn of its reputation before we ever see it. "You've never heard of the *Millennium Falcon*?" says an incredulous Han Solo to Obi-Wan Kenobi in *Star Wars*: Episode IV *A New Hope*. He goes on to brag about its fame and its amazing capabilities, and we know right away that this ship will loom large in the story to come. When at last we do see it, we have all that history in mind, giving the rather junky ship a kind of dignity.

Behemoth of the Imperial fleet, the Star Destroyer conveys notions of power and intimidation in its design. Shaped like a giant arrowhead and bristling with weapons, the destroyer can launch TIE fighters and other assault craft, spreading fear throughout the galaxy.

Pilot visibility reflects another interesting aspect of *Star Wars* spacecraft design. The ships of the Rebel Alliance, for instance, have a visible cockpit, allowing us to see the person in control—a person we just may care about. The Imperial ships, on the other hand, generally have fairly opaque cockpits, making them both more anonymous and more menacing. The TIE fighter's cockpit is a dark octagon, behind which lurks a black-suited pilot. Even more menacing, the droid ships of the prequels (Episodes I-III) have no pilots at all— the ship itself is in control, effectively removing all vestiges of human empathy. Here again, designers need to be mindful of a vehicle's purpose, much like the designers of, say, police cars on our own planet. A police car must exude authority, and yet be approachable as well. After all, the police protect and serve the innocent, in addition to intimidating the bad guys. That's one reason their cars don't have tinted windows; by the same token, it explains the absence of fluffy lettering and pastel colors.

There's nothing fluffy about weaponry in any universe, and *Star Wars* designers took their inspiration for hand weapons from the lethal-looking machine guns, both German and British, of World War II. This effectively grounds them in reality, although the films' weapons fire lasers instead of bullets, putting them securely in the *Star Wars* galaxy. Other considerations dictated the design of the heavy weapons, like the deadly planet-destroying gun of the Death Star and the defensive ion cannon used by the Rebels on Hoth. The Death

Resembling a flying bullet, Darth Vader's modified TIE fighter, center, flanked by a traditional TIE fighter, zooms through the Death Star's equatorial trench in pursuit of Rebel forces.

A British submachine gun from the 1950s, top, and a Broom-handle Mauser from the early 1900s, bottom, served as partial inspiration for the laser-firing blaster rifle and pistols of the *Star Wars* galaxy.

Evolution of a stormtrooper: These fighters began as clones of Jango Fett, commissioned by a mysterious figure. By Episode IV, they will evolve into efficient stormtroopers, serving as infantry for the Empire.

Star weapon's hallmark lies in its being a secret, and thus it's hidden in the bowels of the station, with only a telltale dimple disclosing the emission point. In fact, the station is literally built around the gun, much as an A-10 Warthog is built around its Vulcan cannon in the military arsenal of our world. The ion cannon of Hoth, on the other hand, has a defensive design that enables it to be set up in any location. Its stubby gun protrudes through a protective dome, allowing it to fire from the planet's surface into space; the dome itself can be camouflaged to keep its location secret.

Throughout the saga, in the hands of humans or droids, good guys or bad, lasers remain a defining aspect of *Star Wars* weapons. They're the projectile of choice, whether they appear as short bolts, long cutting beams, or powerful, torpedo-like volleys. They offer up speed, accuracy, concentrated energy, and—let's face it—good old-fashioned star quality. Sometimes, though, the heroes in *Star Wars* outdo Imperial high-tech gadgetry with decidedly low-tech solutions. In *The Empire Strikes Back*, the cable harpoons on the Rebel snowspeeders manage to cripple the menacing Imperial walkers. Ewoks successfully employ stones, logs, and ropes to fell stormtroopers in the *Return of the Jedi*, and Luke uses the Force instead of his computer to direct the shot that blows up the Death Star. In fact, the best weapons aren't always the most explosive or deadly looking ones. Rebel forces—the good guys—often strive for harmony, for a balance between nature and technology, seeking assistance from less technology-reliant cultures. The bad guys, on the other hand, display a correspondingly evil fascination with the power of technology and the weapons it yields. They build armies of droids, huge space stations, and, eventually, clones programmed to do their bidding.

Yet, somehow, these ultimate weapons seem to hold the seeds of their downfall.

Civilian and military designs differ significantly from each other in *Star Wars* , just as they do in our own world. Vehicles and structures aligned with the civilian sector receive a softer aesthetic, courtesy of the Lucasfilm designers, who tend to represent them with colors and textures that are more pleasing to the eye. Military designs, on the other hand, have a form-follows-function aesthetic, with hard angles and flat colors emphasizing might and purpose. The distinction extends even within these categories; the Naboo starfighters of Episode I have a smooth, decorative quality to stress their royal employment, while Episode III's clone turbotank has the no-nonsense angles and bunker-like windows appropriate for a might-makes-right approach. The airspeeders on Coruscant resemble futuristic cars with lots of curved glass and shiny metallic finishes, while the wartime craft are unpainted or dull in finish. And the Alderaan space cruiser in *Revenge of the Sith* displays a diplomatic scheme, with curved and painted panels, decorative lighting, and a well-maintained, glossy finish. We see this same ship more than a decade later in its wartime dress for *A New Hope*, complete with battle-scarred armored plating, retrofitted guns, and all its nonfunctional beauty panels removed.

Droids make up a significant part of the *Star Wars* galaxy, performing tasks that range from translating arcane languages to fighting battles to delivering babies. They come in all shapes and sizes, with C-3PO being the most humanoid in form and character. The design of C-3PO includes a rather friendly face; even without moving parts to show expression, he still manages to emote with simple body language, thanks to Anthony Daniels. Of course, a droid doesn't have to look human in order to possess human character traits. R2-D2 displays distinctly human characteristics through his subtle movements, whistles, and beeps. Battle droids, on the other hand, were designed to stand as poor precursors to the stormtroopers. Their skinny bodies and limbs give them a less menacing impression, and they achieve only limited success in battle. The destroyer droid, next in line, has faster

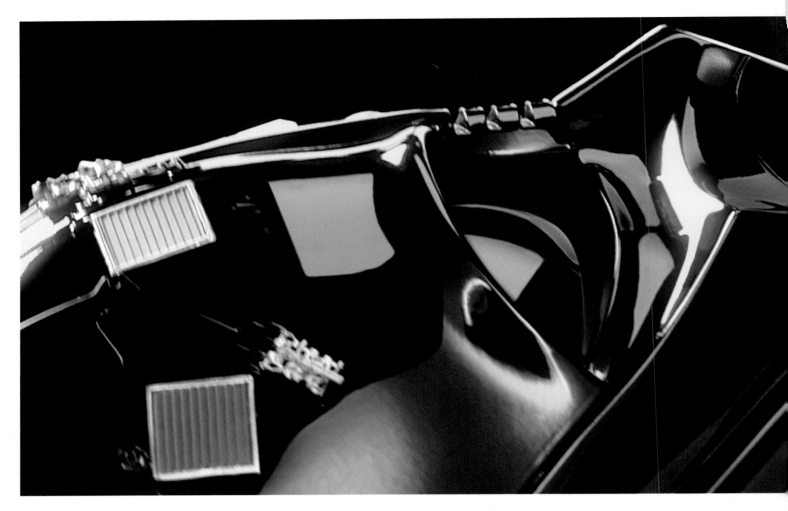

attack speeds, self-shielding abilities, and more powerful weapons. The super battle droid of Episode II represents another upgrade. No attempt was made to give any of these battle droids an individual character; from a design perspective, this lack helps make them appear both formidable and remorseless. A truism maintains that the more a robot mimics humans in its looks and actions, the more disturbing it is to real humans. Keeping a cap on just how human a droid can appear—even with C-3PO—serves to keep the *Star Wars* droids out of the realm of the creepy and within that of trusted companion.

The realm of the creepy, of course, is peopled with half-man, half-machine creations—and, like Darth Vader, the more machine they are, the creepier and more intrinsically evil they become. Indeed, the whole *Star Wars* story revolves around the rise, fall, and redemption of Darth Vader, whose relationship with technology is ambivalent at best. We first meet young Anakin Skywalker as a boy who has an especially strong connection with the Force. Yet starting with his forbidden love of Padmé to his murderous reaction following his mother's death, we see the dark side start to take hold. In *Attack of the Clones*, with the loss of his hand, Anakin begins his transformation into the man-machine known as Darth Vader. But

Vader was not the first of his kind; the Sith had previously experimented with the fusion of flesh and technology in General Grievous, an attempt by Darth Sidious and Count Dooku to create a controllable ally. Grievous is much less organic than he is machine (only his brain, vital internal organs, and spine are left), with a complicated design, while Darth Vader is only about 60 percent mechanical, and his design makes him both more elegant and more mysterious. With all his workings hidden, his appearance leaves the audience in the dark as to how much of him is still human. Those who must face him see only shadow and their own scattered reflection in his glossy black helmet; a blinking red light on his chest and the ominous breathing represent the only detectible technology.

Luke Skywalker—Anakin's son—loses a hand battling with his father, Darth Vader, and receives a mechanical one as a replacement. But, in this case, Luke's technology hides behind synthetic flesh. It serves as a warning to him—he can choose not to follow in his father's footsteps, and instead temper his will in dealing with the Force. Technology itself may not be evil, but the abuse of it, and the seductive power it exudes, seem to represent a defining factor between good and evil in the *Star Wars* galaxy…as could well be said of our own.

In repose, the mask of Darth Vader looks more like a sleek architectural prototype than a design for a villain drained of humanity. Vader's mask serves as the embodiment of dark, faceless evil rather than the likeness of a human being.

LIVING ON CORUSCANT

ED RODLEY

CORUSCANT IS THE BEATING HEART OF THE *STAR WARS* GALAXY, THE CAPITAL OF THE GALACTIC Republic and the Empire that replaces it. This completely urban planet functions as one immense city, the product of millennia of gradual expansion and growth—just like real-world cities, only on an unbelievable scale. Coruscant's skys buzz with endless streams of repulsorlift traffic, and its skyscrapers rise above the clouds and extend down into perpetual shadow— the Republic has paved over every square inch of the planet's surface and built on it. Coruscant has no oceans, no forests to replenish the oxygen in its atmosphere. Almost all of the original wildlife of the planet has been driven into extinction by the relentless advance of civilization. The animals and plants that have survived are those that have adapted to the built environment or arrived accidentally, introduced by the countless travelers who visit the capital. The elimination of its natural environment has turned Coruscant into a kind of giant spacecraft. The planet needs immense orbital mirrors to keep its polar latitudes warm. Atmosphere factories dot the surface, scrubbing the air of carbon dioxide and generating oxygen. A vast network of pipes melts polar ice and distributes it across the planet. As one continuous city, Coruscant has no natural means of waste disposal. Huge amounts of sewage and refuse are blasted into space to burn up in Coruscant's sun. Seen from space, the whole planet glows— the supreme example of technology overwhelming everything around it.

Is Coruscant the city of the future, or a metropolis on steroids? How do scientists and engineers imagine our cities will evolve? This is a pressing question, because by the year 2030 more than half of our planet's people will live in cities, and figuring out how to make future cities more sustainable will become a necessity. Three aspects of urban planning may give us a glimpse of our future: transportation, architecture, and communication networks.

TRANSPORTATION DESIGN

In terms of moving people around, Coruscant looks a lot like our cities. It has personal transportation, mass transit, and long-range transportation. They all hover effortlessly instead of relying on wheels, but they're instantly recognizable as extrapolations of current modes of transport. Will our future be full of trains, planes, and automobiles, or will our children get around in ways we can only imagine? Whatever happened to jetpacks and automated highways? Many of the alternate transportation modes that have been discussed for decades, like flying cars, still linger on drawing boards. Some, like maglev trains, have just begun to come into use. Toyota has debuted a computer-assisted means of parking your car for you. It is almost impossible to see what the next big thing is going to be until it arrives, but some trends are appearing right now.

Transport systems define a city and act as the sinews that hold it together...or keep it segregated. Twentieth-century urban planners, hoping to engineer social behaviors, designed and built "cities of the future" from scratch. One of the more notable failures can be found in Brasilia, the capital of Brazil. Its planners considered automobiles the transportation medium of choice. In the 1950s, they organized the city into single-purpose sectors: governmental, residential, shopping, entertainment, and so on. As a result, Brasilia has none of the rich interplay of people and commerce that exists in a more organic city. It is easy enough to drive around, but you have to drive to do anything. Building a city for cars didn't work out very well.

With nowhere to go but up, the builders of Coruscant have turned their city-planet into a metallic sea of skyscrapers; most real-world urban planners —and dwellers—would bemoan the total lack of green space.

It seems clear that the end of the gasoline-powered automobile is in sight. We can no longer take cheap, plentiful oil for granted. Even major manufacturers, who have invested billions of dollars in producing vehicles that run on oil, have begun to design hybrids that combine gasoline and electric motors. In the U.S. and abroad, pundits and policy makers now tout the potential of ethanol, or of the hydrogen economy that uses hydrogen fuel cells to generate electricity, not just for cars but for all the powered devices we seem increasingly to need in our daily lives.

More immediately, the problem of congestion will provide the stimulus for fundamental changes in how we get around. Modern cities are not managing transport capacity well enough to keep pace with the numbers of people trying to move around in them. Congested periods last longer and delay more travelers and goods than ever before, translating into more precious fuel consumed by vehicles idling in stalled traffic. And if that isn't bad enough, we're facing a global trend toward urbanization. In the developing world, people are moving into cities at an unprecedented rate. Over the next decade, more than half a billion people will make the move from rural to urban areas, and they'll all need to get around. The car-based transportation system we are used to cannot accommodate that kind of growth. There is nothing like a perceived threat to encourage people to come up with new ideas.

Short-term solutions include gas/electric hybrid cars and ambitious attempts to design vehicles powered exclusively by solar or fuel-cell energy. Urban planners have consciously moved away from the 20th-century model of the city core with its suburban sprawl. Advocates of so-called smart growth envision cities as urban villages where people can access the goods and services they need without a car. U.S. and European lawmakers have begun changing zoning laws and creating vehicle-free zones to encourage foot traffic; larger cities like London even levy special taxes on drivers in congested parts of the city. Advocates of mass transit continue to push the speed limits of moving large numbers of people, and high-tech train technologies like maglev promise to bring fast and quiet mass transit to new heights.

ARCHITECTURE

Our innovative buildings don't look much like the Coruscant cityscapes. New "green" buildings often sprout grass on their roofs to capture and collect rain water. Their internal heating and air conditioning systems are designed for maximum efficiency of resource usage. They are designed to funnel natural light as much as possible—it's free, after all, so why not use it? The U.S. Green Building Council now promotes a Leadership in Energy and Environmental Design (LEED) Green Building Rating System, with the hope that market-driven ratings will accelerate the development of green building practices; judging by the number of builders applying to have their designs certified, LEED can already claim some success.

A number of urban planners advocate a vision of future cities that try to achieve high population density along with the intimacy of a small village, where residents can get everything they need without going far. These cities look very different from Coruscant. Built on a much

Do elevated highways and acres of high rises provide high-quality urban life? That's the approach of many cities, including densely populated Tokyo, Japan.

smaller scale, they are microcosms of a traditional city and can be replicated like clusters of fruit on a tree. In densely populated Japan, designs for future cities come closest to the *Star Wars* model; they include mega-skyscrapers reaching a mile into the sky and housing half a million people, and giant pyramids containing entire cities inside their skeletal frames.

But even these concepts differ from the fantasy. The amount of green space they incorporate into the fabric of the city is much larger than in Coruscant, or traditional cities here on Earth, for that matter. These new urban environments also tend to be much more open to the weather—a rejection of the "sealed" building where the air must be artificially circulated. The city of Chicago, Illinois has embarked on an ambitious plan to become the most environmentally friendly city in the world, embracing new concepts that range from sustainable urban growth and green building projects to modifications of existing buildings and infrastructure.

COMMUNICATION NETWORKS

Coruscant and the real world are pretty much aligned in their communication networks—you can't see them, but they're there. We live in an age when these networks are becoming more and more integral not only to interpersonal communication, but to almost any aspect of modern society. The extensive penetration of computer processing into manufactured goods has created a demand for communication that far exceeds people talking to one another. Now, businesses and governments rely on data communication networks. Home computers have gone from being number-crunching machines to communication terminals. We carry connections to wireless communication networks in our pockets and purses. By and large, these networks remain invisible; they exist in the ether and behind the walls, but they are rapidly becoming indispensable. We are moving from an age of solid infrastructure to one where lots of devices offer swift communication without having a cord coming out of their back.

Our homes are becoming the hubs of digital networks, and trends seem to indicate that the amount of communications will only increase. Your computer may communicate with other computers, but soon your house may communicate with you. Researchers in several countries are trying to imagine how to make "smarter" houses that can actually learn to anticipate our needs. Someday, your living room may be able to tell if you're cold and turn up the heat without your asking, or turn on the lights when you start reading and turn them off when you leave. Look around your own home and imagine the number of things that might conceivably incorporate communication systems. Doors, windows, heating and lighting controls, even furniture might become parts of home networks.

All over the planet, architects, city planners, governments, and ordinary people are imagining the city of tomorrow. It seems likely that our future cities will not resemble Coruscant much, but what will they look like? The best response to that question might be another question: What do we *want* our cities to look like? The answers to that question will lead us to a future unlike anything we've seen—a future that promises to be every bit as exciting as the movies.

PART TWO

ROBOTS AND PEOPLE

The Future of Robotics

THE FUTURE OF ROBOTICS

ED RODLEY

ROBOTS HAVE BEEN MAINSTAYS OF SCIENCE FICTION AND FANTASY LITERATURE SINCE THE 1920s. In fact, they've been such memorable characters that many of them are household names—Robby the Robot, HAL 9000, the Terminator. The *Star Wars* films abound with robots—good, bad, and indifferent—including the two most famous fictional robots in the world, R2-D2 and C-3PO. George Lucas conceived the two droids (a shortened form of "android," which *Star Wars* introduced into common usage) as the audience's guides to the *Star Wars* galaxy. They play a central role, taking part in all the momentous events engulfing the Skywalker family. The two share the honor of being the only characters to survive all six films. Their foot- (and wheel-) prints are immortalized in front of Grauman's Chinese Theatre in Los Angeles where *Star Wars* first opened in 1977; a June, 2005 web search for "R2-D2" and "C-3PO" came up with over 1,200,000 hits. Not bad for a motorized trash can and a guy in a gold suit.

Robotics is a unique field of endeavor. No other technology ever proposed or developed has posed such a direct challenge to humans, or been so predefined by fiction. On page and screen, we've all seen the promise and perils of robotics played out countless times, often with dire consequences. By the 1940s, robots had already become stock villains in pulp fiction. About this time, Isaac Asimov began writing science fiction that imagined robots as tools developed by engineers, rather than the creations of madmen bent on destroying humanity. He proposed three laws of robotics (a word he coined) that would govern the behavior of his fictional creations. First, a robot may not injure a human, or allow a human to be harmed through inaction; second, a robot must obey humans except in violation of the first law; and third, a robot must protect itself except in conflict with the first two laws. Asimov's robots had complicated relationships with humans, and the human characters had mixed feelings about sharing their world with robots; these themes continue to play out in popular culture.

The impact of cultural forces on technology is apparent in the way robotics has developed in the U.S. as opposed to Japan. American authors and filmmakers gave us a string of robotic villains, but Japanese popular culture was dominated for decades by one robot—*Tetsuwan Atom*—Mighty Atom, or Astro Boy, as Americans called him.

In both evil and benevolent roles, mechanical actors like Robby the Robot have shaped how the public regards robotic technology and artificial intelligence. Robby plays ball with Richard Eyer, his co-star in the 1957 film, *The Invisible Boy.*

From his first appearance in 1951, Astro Boy left his mark on Japan. Looking like a small boy, he possessed a soul and feelings; he was unfailingly helpful, and dedicated to protecting his friends and Earth from endless threats. Astro Boy's "birthday" on April 7, 2003 was a major event in Japan. If you ask Japanese roboticists how they want to be able to interact with their robots in the future, many will say "not as appliances or tools, but as friends."

This unabashedly positive view of robots has combined with another cultural trend to drive billions of research dollars into robotics in Japan. The Japanese have one of the lowest birth rates in the world, and an aging population. The net impact is that there soon will be more older than younger Japanese. The elderly have traditionally enjoyed high status in Japanese society, but who will take care of all these elders when they out-number their children? The Japanese prefer not to employ foreign laborers, so they won't just import workers. The answer? Humanoid household robots, of course. This is why robotics research is being conducted at the university level and also in corporate research laboratories. Companies like Sony, Toyota, and Fujitsu all have active robot labs, and robots like Honda's ASIMO demonstrate the potential of humanoid robots to live and work among us. Still, we have a long road ahead.

So why don't we have robots that approach the functionality of C-3PO or even R2-D2? Tremendous advances have been made in the past 50 years in robotic mobility and sensing, and most importantly in computer-processing power. We have sent robotic rovers to other planets; built robots capable of surpassing human senses like vision, hearing, and smell; and designed robots that can outthink humans in certain limited domains, such as playing chess. Until very recently, though, we have not been able to integrate all these capabilities into one reasonably sized package that is smart enough to understand spoken instructions, nimble enough to handle obstacles like doors and stairs, and powerful enough to run longer than a few hours at a time. Modern robotics presents a gigantic integration problem—how to perfect and coordinate all the differ-ent systems that have to work together so that a single robot can rival R2-D2 in mobil-ity, perception, and cognition.

If we are on the verge of a Golden Age of Robotics, as Rodney Brooks suggests, how we treat robots and how they treat us will leave the realm of the hypothetical and enter reality. And that reality may not look much like *Star Wars*. The kinds of robots on the drawing boards today may live in our houses, take care of our elders, and be more emotionally astute than poor C-3PO, who never manages to under-stand sarcasm. Our future robots may not even be mechanical constructs, but rather biomechanical creatures with muscles and skin. Truth may well be much stranger than fiction.

Robots have already started to enter our lives, and as they grow more sophisti-cated, they will doubtless wind up being used in unexpected ways. Computers were originally developed as specialized military equipment. The thought that they would become a mass-communication medium like the Internet, and a platform for inter-active games, probably never occurred to any of the engineers who worked on them in the 1940s. We are just beginning to see signs of the same thing with robots: From one-of-a-kind research tools, robots have become toys like AIBO and household appliances like the Roomba. Illah Nourbakhsh describes some of the challenges of designing lunar habitats for the next wave of astronauts who will go to the moon, and how NASA may use robots to create an entire moon base before the first human even arrives.

One of the most exciting fields of robotics research is the crossover between robot-ics and prosthetics. If we can create limbs that robots can manipulate, why not make ones for a person to use? The results could bring enormous benefits...but the combi-nation of biological and mechanical parts troubles many, and conjures up the specter of the ultimate man-machine nightmare—Darth Vader. For George Lucas the filmmak-er, Darth Vader's menacing black life-support suit and prosthetic limbs provide visual cues that he has lost some of his humanity. Does incorporating technology into our bodies somehow make us less human? And what counts as technology? Contact lens-es? Does an artificial heart or limb count, or an organ grown in a lab? Richard Satava touches on some of the incredible medical implants of the past decade, ranging from the microscopic to entire limbs and organs.

Here's an irony of robotics research today—while it's relatively simple to get robots to perform various complicated tasks, it remains almost impossible—so far—to create robots that can do many of the things a two year-old child does with ease, such as getting up after a fall. Much of Cynthia Breazeal's research has to do with building robots that can not only learn, but also express themselves in ways that humans instinctively understand. Her first sociable robot, Kismet, spoke gib-berish, but otherwise responded appropriately to conversational cues. If you yelled at Kismet, it acted hurt; if you got too close, Kismet pulled back from you. Research in sociable robotics aims to lessen the communication burden on humans, so that we'll ultimately be able to interact with robots in the same ways we com-municate with one another.

In the *Star Wars* saga, C-3PO manages to convey abundant emotional information simply through his voice and gestures, even though this behavior has no immediate ben-efit to C-3PO. Humans are emotional beings and we expect an emotional response in our communication partners.

Roboticists are helping transform robots from tools to partners, from equip-ment to coworkers. Will we one day think of them as friends? We can only imagine what the robots of tomorrow will look and act like, but it seems clear that robots are here to stay.

ARTIFICIAL INTELLIGENCE

RODNEY BROOKS WITH ROBERT NAEYE

RODNEY BROOKS, Panasonic Professor of Robotics, and Director, Computer Science and Artificial Intelligence Laboratory, Massachusetts Institute of Technology; co-founder and chief technical officer of iRobot Corp. Dr. Brooks specializes in the engineering of robotic intelligence, and in the applications of human intelligence to the construction of humanoid robots.

ROBERT NAEYE, Senior Editor, *Sky & Telescope*. Naeye has authored two books and recently received an award for science journalism from the American Astronomical Society.

I HAD JUST BEGUN WORK AS A GRADUATE STUDENT IN STANFORD UNIVERSITY'S ARTIFICIAL intelligence laboratory when the first *Star Wars* movie came out in 1977. I went to see it and thought it was great. The technologies portrayed in the film confirmed my belief that there were many interesting things scientists could do in the future to build intelligent robots. I had already been inspired by an earlier movie, *2001: A Space Odyssey,* and *Star Wars* fit right in. I was particularly fond of R2-D2. Not only was this droid cute and friendly, but it also communicated without using words, something I found fascinating. I never understood how it managed to go up the stairs between scenes, though, and stairs remain a problem for robots. A three-pound dog can run up and down stairs, but even in 2005 we have yet to figure out how to build a robot that can manage such a seemingly simple task.

What, exactly, is a robot? Different people have different definitions, but I would say that a robot is a machine with sensors; it senses the world around it, performs computations, and then engages in physical actions outside of its own body. Without these external actions, many types of machines, such as dishwashers, conceivably could be called robots.

I believe we're living in a golden age of robotics, a time of transition from a pre-robotic world to a world in which robots are a simple fact of life. A century ago, people suddenly found themselves faced with a similar transition, from having no electricity to living in a world that was fast becoming thoroughly electrified. In this century, many children have enjoyed robot toys for several years. Now robotic appliances are starting to enter our homes, with devices like robotic vacuum cleaners and lawn mowers. Robotic technology is being introduced in automobiles, like the Toyota car that parks itself automatically with the press of a button. This technology will be incorporated into a variety of machines that we use daily, with the addition of components that sense the world, compute accordingly, and then perform appropriate physical actions. A familiar example of this technology can be found in the anti-lock braking systems in certain automobiles. Within half a century or so, robots may be everywhere in our lives in the same way that electricity is everywhere today.

Imagine going back in time 25 years to 1980, and telling a typical American family that everyone in 2005 would have computers in their kitchens. They would think you were nuts. Hollywood's take on computers back then was a big box that spits out magnetic tapes.

Star Wars droid R2-D2 might resemble a trash can with legs, but the character's loyalty, coupled with endearing beeps, whistles, and gyrations, helped create the power of personality.

Putting one of those contraptions in a kitchen made no sense. But thanks to miniaturization, many people now have computers in kitchen appliances, from microwave ovens to refrigerators. Because it's so prevalent, people often don't realize they have this computer technology all over their homes. In the future, I believe robotic technology will be deeply embedded in our lives, but it won't be Hollywood's current version of the robot. A century from now, robots will perform services that we can scarcely imagine today.

At the Massachusetts Institute of Technology Computer Science and Artificial Intelligence Laboratory (MIT/CSAIL), my colleagues and I strive to build robots that can work among us and, continuing the long history of technology, make our lives easier. The challenges are fourfold: achieving the visual competency of a two-year-old, the manual dexterity of a six-year-old, the ability to move around in peoples' homes, and the ability to interact socially so that people can enjoy a robot's presence and easily predict its behavior.

Some of these challenges are easier to address than others, and working with robots has taught me a great deal about what it means to be human. Many tasks that are simple for humans turn out to be extremely hard for robots. Even things that a two-year-old child can do, such as recognizing objects, tend to be very difficult for robots to accomplish. The dexterous manipulations of a six-year-old child are extremely difficult for robots. I have already mentioned the problem with stairs. Social interactions, on the other hand, such as nodding when you see someone and making eye contact, turn out to be much easier for robots than anyone expected.

I realized the importance of social cues when we built our first humanoid robot, Cog, in the mid-1990s. When it looked quickly from side to side, it had an inanimate appearance. But when we gave it the ability to swivel its head while keeping its eyes locked onto an object, people suddenly felt it was alive. That's a fairly simple thing to program into a robot, but it triggered a reaction in everyone who saw it, out of all proportion to its level of difficulty. We almost felt as if Cog was not a machine anymore, but rather a living being.

One of our later robots, Kismet, pushed even harder at the frontiers of social interaction. Kismet has now been retired to the MIT museum, but we made great strides as we developed it. Not only would Kismet make and break eye contact, it would offer subtle verbal cues even if they had no actual meaning. It would respond to a person speaking, as we often do when we say, "Uh-huh, yeah." All of those little things made people feel as if they were carrying on a conversation with Kismet. Even when Kismet didn't understand what a person was saying, the inadvertent social cues that the speaker provided, and to which Kismet responded, were enough to convey the impression it was paying attention.

For robots to be considered true "friends," we will need to develop ongoing relationships with them. Robots like Cog and Kismet started to achieve relationships with people by engaging us, maintaining eye contact, and interacting socially for about a minute. But until we can extend that time period considerably, people will not attain the level of relationships with robots that they can have with cats or dogs. We will have to push robotic technology

About the size of a volleyball, the PSA—Personal Satellite Assistant—can navigate the microgravity environment of the Space Station. NASA designed its clusters of sensors and thrusters to function as an extra set of eyes and ears.

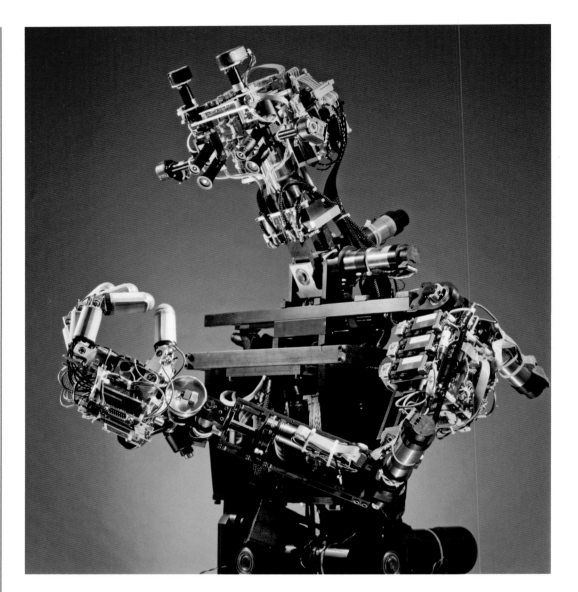

Developed in the 1990s by MIT's Rodney Brooks, Cog is an upper-body, humanoid robot designed to replicate how humans sense and interact with their world. Knittting various sub-disciplines in robotics, such as sight and touch, into a functional whole, Cog has binocular vision, and arms and hands with the same range of motion as human joints. Working with Cog has helped researchers understand interrelationships of robotic systems, and how robots relate to humans and their environment.

Biomolecules: Organic compounds that naturally occur in all living organisms, primarily composed of carbon and hydrogen atoms. Examples include vitamins, proteins, and DNA.

a lot further, so that our robots can attain a sense of self that operates over months and years, before we can form any deep relationships with them.

Can a machine ever possess free will or self-awareness? If a robot has a computer for a brain that is run by software, how can it have free will? Well, the same question can be extended to humans. How can people have free will if they can be viewed as collections of biomolecules interacting according to known physical processes? For example, at one level I see my children as big bags of biomolecules. But at another level, I love and adore my children, and I interact with them on a very personal basis. I view them as living, thinking beings who undoubtedly possess free will. Human beings manage to interact with one another and have a belief in free will, and that belief is quite valid because it allows us to change what we decide to do in various situations. If we believe that people have free will, we should believe that future robots can possess free will too.

What about true feelings—can robots have them? When scientists talk about DNA, we don't talk about how one biomolecule links up with another, and then the soul intervenes and makes the connection. Scientists speak in terms of physical forces, attractions, and repulsions. There's no magic. But our feelings arise from those interactions, and there's no denying that we have feelings. So by understanding all these interactions, we should be able to build machines that have feelings. Whether we're smart enough to build such machines remains to be seen, but in principle it should not be viewed as impossible.

The possibility that robots can have feelings and free will frightens some people, because they think these things are

> If a robot has a computer for a brain that is run by software, how can it have free will?...How can people have free will if they can be viewed as collections of biomolecules interacting according to known physical processes?

for humans only. But over millennia scientific progress has forced humanity to retreat from the idea that we are special. Five centuries ago, Nicolaus Copernicus showed that the Earth revolves around the sun, instead of the other way around. Suddenly, the Earth was no longer the center of the universe. With Charles Darwin's theory of evolution, we lost our special sense of being different from the animals. We now know that we are made of the same stuff as plants and animals and that we share a common ancestry. But in recent times, reactions to this fact have reverberated around the U.S., where some people prefer to deny reality rather than accept that evolution takes place and that our origin was part of that process.

To help maintain our sense of having a special nature, IBM conceived the slogan, "Computers don't think; people think," when it developed better computers in the 1960s. IBM wanted people to feel computers were different from us. As robots' capabilities improve, people will worry about losing some of our sense of being special and about having to share that with machines. We're seeing an expression of this anxiety with the restrictions placed on stem-cell research, an anxiety which comes partly from the revulsion that human flesh can be subjected to technological manipulation.

People sometimes ask me if a robot can have a soul. It might not happen for twenty or even fifty years, but ultimately we will start ascribing souls and personhood to certain robots with whom we develop ongoing relationships. And then they will have souls, and even self-awareness, in the same sense that you and I do. That is why I treat people as being much

Surrounded by droids in violation of the real-world "don't harm the humans" rule of robotics, Anakin, Padmé, Mace Windu, and an assortment of Jedi are outnumbered but not outsmarted.

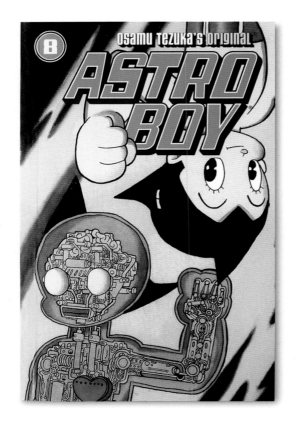

Boy scout of the robot world, adorable Astro Boy insinuated himself into Japanese culture for more than five decades, and guaranteed that Japan would view robots as friendly helpers.

more important than mere collections of biomolecules. Discussions about free will, feelings, and self-awareness in robots inevitably lead to concerns about whether we're planting the seeds of our own destruction, a common theme in science fiction.

Hollywood is in the entertainment business, and one form of entertainment is to play to our fears, including the commonly held fear that after we create, nurture, and teach robots, they will start to think they are smarter than we are, and bad robots will want to push humans out of the way in order to take over the world. Will we someday see armies of evil battle droids, like those in *Star Wars* Episodes I, II, and III, or the nightmarish futures depicted in *The Matrix* or *The Terminator* films? Reality almost always turns out to be quite different from how Hollywood imagines it. For the foreseeable future, we're not going to build dangerous robots any more than we intentionally build dangerous trains that go out and kill people. If robots start to misbehave, we'll change them. Nobody will suddenly build a really crazy robot that can do everything all at once. There would have to be a slightly crazy robot before that, and a mildly crazy robot before that, and so on, back to the present. We will control and maintain a comfort level with the types of robots we build.

Another concern involves people with bad intentions designing robots to do terrible things, just as terrorists today build improvised explosive devices. But building an intelligent robot that goes out of control and starts killing people in large numbers would be sort of like some guy coming into the kitchen one day and saying to his wife, "Martha, I accidentally built a 747 in our backyard." It might happen in a Hollywood movie, but it's not going to happen like that in real life.

It's dangerous to make sweeping generalizations about entire societies, but the Japanese often seem more comfortable with robots than Americans. Japan has a long history of being interested in robotics, and the beloved Japanese robot character named Astro Boy has been a force for good for more than 50 years, helping shape a tendency in Japan to think of robots as beneficial. Americans tend to be more afraid of robots and the bad things that could happen because of them. These cultural perceptions are reflected in advertising. Japanese ads are often much more overtly technological. For example, a television ad will point out a washing machine that has artificial-intelligence technology built into it. In contrast, we rarely see that kind of advertising in the U.S..

But one common fact links Japan and America, as well as Europe and South Korea: changing demographics. Our populations are aging, with the elderly increasingly outnumbering the young. Scientists looking for ways in which robots can take part in elder care must consider different societies and their different views of how people and robots should interact. In the United States, we tend to think of robots as tools that just do something useful on the side. But much of the research in Japan deals with robots as companions, an uncomfortable concept for many Americans.

Some people think of a robot as being like a car that is controlled by a driver. But robots will be more independent in the future. We'll continually need to adjust their sensing capabilities so they do what we want them to do, while rejecting actions we find dangerous or annoying. There will be a constant back-and-forth communication between scientists who build robots and consumers who use them. The scientists will respond to the consumers by tuning robotic capabilities to maintain a certain comfort level.

Throughout history, new technologies have raised new legal questions. For a long time, we had to know whom to hold legally responsible when a horse damaged property or did something bad. A century ago, it was usually the horse's owner who had to answer for the horse's actions. In the foreseeable future, robot owners will be held responsible for their robot's behavior. Knowing this, people will buy robots only if they have a high degree of confidence in how they will operate. But as the field advances and robots develop more initiative, the legal questions are going to become harder

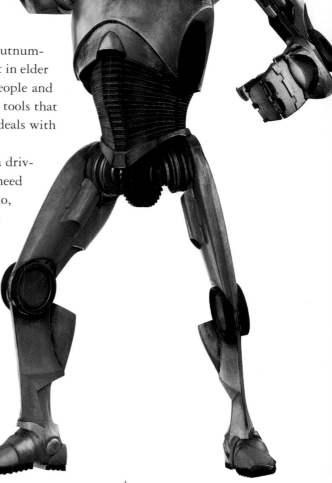

A featureless recessed head and built-in wrist blasters proclaim the belligerent intent of the super battle droid, designed with only one purpose in mind: to vanquish its foes.

An unmanned aerial vehicle (UAV), the Predator flies above an aircraft carrier during a simulated Navy reconnaissance flight in 1995. UAVs perform reconnaissance, combat, and support roles that previously could be handled only by piloted aircraft. Predators have logged more than 65,000 hours of flight time, half of them in combat.

to resolve. Issues will arise that have never been tested before; both the issues and their ultimate resolution are difficult to predict.

Robots have had, and will continue to have, a positive impact on the economy by making factories more efficient and by doing jobs that people don't want to do, such as cleaning up chemical waste-disposal sites. Understandably, people are concerned that robots will take all of our jobs. But it's important to remember that people once said the same thing about computers. Now we have more computers than ever, yet we're working harder than before and there are still plenty of jobs. That's how it's going to be with robots, too.

Many people also express concern about the military role of robots, but they ignore the fact that using robots for certain dangerous missions can save countless lives. For example, machines are currently being used to detonate homemade bombs in Iraq, so American soldiers aren't blown to bits trying to disarm them. A lot of research has enabled robots to clear minefields, which has also saved lives. The United States Armed Forces will continue to use robots as military aircraft, such as the reconnaissance drones flying over Iraq. Robots will change how war is waged in the future, just as new technologies have continually changed the face of war over the centuries. Military applications will drive robotic technology forward, and the benefits from that research will filter down for peaceful uses in society at large.

Until very recently, living beings and machines have existed within two separate domains. But now, fifty years after the study of molecular biology began, scientists are starting to engineer biomolecules and living systems, and integrating nonbiological structures into systems such as our own bodies. This is often driven by medical needs; for instance, scientists have developed cochlear implants for deaf people, retinal implants for the blind, and even robotic limbs for amputees. With its hyperdrives and giant space cruisers, the *Star Wars* saga depicts certain levels of technology far ahead of our own, but we are already seeing the forerunners of the robotic limbs given to Luke Skywalker in Episode V and Anakin Skywalker in Episodes II and III. In the future, I expect to see more robotic technology implanted inside people's bodies to replace lost abilities, or to augment abilities as people age.

Today we build robots made of silicon and steel, and, not surprisingly, we think of them as machines. In a sense, people are also machines—but we're biochemical machines, made of biomolecules that interact in many different ways as permitted by the laws of physics. Twenty to fifty years from now, we will likely build our robots out of similar biological material. In essence, we'll make biological robots that won't look anything like R2-D2 or C-3PO. The challenge for us is to attain the same level of complexity found in our own biomolecules, so that robots can be as adaptable as we are.

Today, for example, we can grow a tree, cut it down, and then build a table from its wood. Laboratories are pioneering a new field, called synthetic biology, to control the proteins that form inside living cells. Someday, scientists will be able to build a table by growing it directly from biological material. In fact, in thirty or forty years, we'll not only grow structures directly from such material, we'll also put neurons, computers, and sensors into machines fabricated from living material. This technology will eventually allow scientists to build new, stand-alone robotic systems made out of synthetic biological substances. We're already working toward this future in our lab at MIT.

To their credit, the *Star Wars* filmmakers envisioned the capabilities of truly intelligent robots, but they were unable to anticipate this interaction of biology and robotics. These two fields are going to merge over the next fifty years in exciting ways that even the most visionary practitioners of science fiction have not foreseen. *Star Wars* creator George Lucas showed us a universe where droids could do amazing things, from translating millions of languages to repairing damaged spacecraft on the fly. But much of the robotic technology depicted in *Star Wars* was actually quite low-tech even by contemporary human standards. The movies gave us a vision of highly mechanistic contraptions made of metal and wires. They showed us some of what robots might be capable of doing in the future, but they did not show us what future robots will look like.

I will have succeeded in my life's quest if we reach the point when graduate students feel badly about switching off a robot we've built, because switching it off would be tantamount to killing it. I can't imagine switching off any of my children and letting them linger in suspended animation—that would be horrible. If we can make a robot that causes us to feel the same way about switching it off, that would mean that we had built a living creature, even if it was made of silicon and steel instead of biomolecules. Over the next few years we'll see whether robots really will remain creatures of silicon and steel, or whether they too will become biological entities.

Hazardous duty becomes slightly less hazardous thanks to a remote-controlled PackBot, deployed in Iraq in 2005 by a U.S. soldier. This Explosive Ordnance Disposal robot helps soldiers dispose of bombs while keeping a safe distance.

Neurons: Grayish or reddish granular cells that are fundamental functional units of nervous tissue.

MAN, MACHINE, ROBOT, AND CYBORG

HOW FAR FROM REALITY?

RICHARD M. SATAVA

RICHARD M. SATAVA, MD, FACS, Professor of Surgery, University of Washington Medical Center, and Program Manager, Defense Advanced Research Projects Agency (DARPA). A practicing surgeon, Dr. Satava has special interests in military surgery, telepresence surgery, and the role of surgery in space.

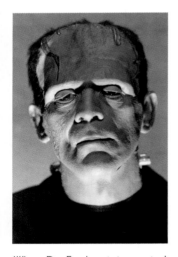

When Dr. Frankenstein created his monster in the 1931 film, its misshapen skull, sutures, and protruding joints proclaimed a tragic lack of humanity.

AS A SPECIES, WE'VE LONG BEEN FASCINATED WITH ROBOTS AND MACHINES, ESPECIALLY intelligent machines that resemble … well, us. Although Mary Shelley created the first synthetic person in her 1818 novel, *Frankenstein*, it was the Czech writer Karel Capek who first used the term "robot" in 1920, describing a machine designed to take over repetitive, tedious factory jobs, and introducing the concept of a machine that might ultimately think and behave like a human. Novels such as Aldous Huxley's *Brave New World* and George Orwell's *1984* brought legitimacy to speculation about future societies in which people were turned into nonthinking automata, but the prospect of an intelligent robot or synthetic human remained in the world of fantasy.

In the 1950s, with the rising popularity of authors like Isaac Asimov, Arthur C. Clarke, and Phillip K. Dick, the idea of intelligent robots entered popular culture and became a fashionable topic for discussion—though admittedly a disparaged one in scientific and intellectual circles. The movie *Forbidden Planet* (1956) featured one of the first robots to assist humans, but even Robby the Robot did not attempt to actually look human, and no self-respecting scientist took robotics seriously. Meanwhile, however, the business community continued to use machines of manufacture, which became more and more automatic, finally acquiring the term "factory robot." Yet these robots did not bear any significant resemblance to humans; far from being anthropomorphic, they weren't even complete—they usually took the form of an arm, or video system with conveyor belt. But their efficiency could not be denied, nor could the animosity they stirred up among those with fears about job protection.

Then, in 1977, a remarkable movie called *Star Wars* propelled into the human consciousness the idea of robots as friendly and intelligent creatures, and added the term "droid" to the popular-culture vocabulary. Cute little R2-D2 and a very insecure C-3P0 brought a brand-new sense of humanness to what had previously been viewed as cold, dark, and menacing. And while the possibility of a humanoid robot was well beyond the science of the times, this did not prevent the public—already used to automobiles, computers, and factory robots—from embracing a future likelihood of intelligent and sometimes human-like machines. It didn't matter that the film only simulated robots, and that actors were giving them their human attributes—the seed had been planted. Other films portrayed robots

Sinking deeper in thrall to the dark side of the Force, Anakin clasps his mechanical hand—in the films, a symbol of his failing humanity—to an increasingly defiant heart.

At England's Papworth Hospital, surgeons insert an artificial valve into a patient's chest during an open-heart procedure—a clear case of technology offering humans a new lease on life.

Cyborg: A bionic human; that is, a human who has had organic tissue replaced, combined with, or augmented by automatically controlled, mechanical-electrical systems.

as indistinguishable from humans, and a number have used special effects to create a creature which is half human and half machine—the cyborg. And then, of course, there was Darth Vader, cloaked all in black and projecting that menacing voice from the robot-like face guard he always wears. The ultimate *Star Wars* villain, he's described by Obi-Wan Kenobi as "more machine now than man," and this lack of humanity makes him all the more compelling. Until Episode III lifts the veil, we're left to wonder just how much of him has been replaced, and how, and why.

Recent scientific success in the area of artificial limbs has begun giving credibility to the possible, some say inevitable, creation of a combination man and machine, though we can hope and expect the result will be something other than Darth Vader. In fact, the basic concept of replacing human organs or structures with electronic or mechanical devices—witness the knee and hip replacements performed routinely in hospitals all over the world—raises the question of cyborgs. Incredible advances in healthcare technologies, such as implanted automatic pacemakers, metallic prosthetic joints, synthetic heart valves, and replacement hearts, give credibility to a time in the future when most any part of the human body will be replaceable, as happens to some *Star Wars* characters, including Luke Skywalker and Darth Vader, and in the television series *The Six Million Dollar Man*. Not only will body

parts be replaced, but some will be transplanted from humans or animals (nearly all major organs are routinely transplanted today), while others may soon be synthetically grown, creating a person whose body is partly synthetic and partly natural in its biological makeup.

In spite of all this progress, there remains a determined and unfaltering belief that some uniquely human quality exists that will transcend any thought of replacing humans with totally intelligent, and possibly biologically grown, robots. Factors such as intelligence, emotion, personality, creativity, faith, and theology are all invoked as proof positive that more than simple scientific replication is embodied in what it means to be human. Of course, only some of the questions that arise can be answered scientifically.

A number of basic definitions and premises must be made in order to focus the answer away from the spiritual, which cannot be physically measured, and toward those scientific accomplishments and aspects which can be quantified. Science is moving forward through careful analysis of the physical characteristics of humans that can be measured and understood, and then making slow, steady progress in mimicking the functions of these organs (for example, legs, or hearts) and processes (locomotion, or blood circulation) with engineered parts, chemicals, or information-based systems. In certain areas, the results are astounding; in others, we are just scratching the surface.

The overwhelming challenge of robotic vision has been three-dimensional sight. Devices like the tiny 3-Dimensional Artifical Neural Network (3-DANN) could allow robots to pick out items against a visually cluttered background and successfully process the data.

First U.S. service member to be fitted with an Otto Bock C-Leg, Lt. Col. Andrew Lourake is a special missions pilot who has returned to the cockpit following above-the-knee amputation, thanks to the C-Leg's abilities: It's sophisticated microprocessors enable it to react to the wearer's movements and make adjustments 50 times per second, based on stride, speed, and walking surface.

Preceding the complex issue of creating an entire humanoid robot with multiple, inter-operable systems, significant steps have occurred in developing human replacement parts and individual organs. The simplest of these are joints shaped to resemble the parts they're replacing, such as hips or knees; made of titanium, they're inert and nonresponsive. However, new prostheses have been developed with tiny sensors and actuators, moving parts made of micro-electromechanical systems (MEMS), which are microscopic machines etched into silicone in the same way computer chips are etched. These devices sense motion, and can change and adapt to their environment. The Rheo Bionic Knee, developed in Iceland, and the C-Leg Microprocessor Knee, developed in Minneapolis, Minnesota, have the capacity to provide sensory feedback and active control. MEMS are also being placed into the instruments of surgical robots, in order to give the surgeons using them the sense of touch.

In ophthalmology, researchers have developed a very simple and crude artificial retina for people who have lost their sight but still have intact nerves. Implanting a very thin chip, similar to the tiny chip camera used in cell phones, allows light to be transmitted back to the nerves of the eye, so that a very fuzzy image can be seen by the patient. Although well below the level of normal vision, the ability to see a shadowy representation of the world becomes an important first step for those with no vision at all.

In the 1990s, researchers developed a treatment for Parkinson's Disease by identifying the place in the brain where a focus of neurons spontaneously fires independently, causing the typical tremor of Parkinson's. A tiny wire or probe implanted in this specific area of the brain could then be activated, preventing the nerves from firing and thereby suppressing the tremor. A number of researchers, including John Donahue at Brown University, have taken the next step—implanting an array of probes into the brain of a monkey, recording the brain's activity with the purpose of interpreting the signals that control movement. A brain chip with 100 probes, called BrainGate, is implanted into the monkey's motor cortex, where it reads the signals of 100 different nerves and sends them to a computer for interpretation.

Using a video monitor and a joystick, the monkey learns to move a red ball to sit on top of a green ball; each time the red and green ball touch, a robot arm swings over and feeds the monkey a small treat. When scientists disconnect the wire from the monkey's brain to the computer and reconnect the wire directly to the robot arm, it takes the monkey a couple of weeks to realize it can move the robot arm to feed itself without moving the joystick. A number of monkeys can now move a robot arm simply by thinking.

Recently, one of these chips has been implanted into a quadriplegic patient, and now this patient can control a television, move a cursor on a computer, give simple commands to a robot arm—all by simply thinking. In the next year or two, this patient may well be able to position the robot arm with precision, picking things up or even feeding himself. These, of course, are the simplest of tasks, and the effort to conquer them illustrates just how much more we need to do to provide real human abilities; nonetheless, this represents an extraordinary step toward implantable devices that could control artificial—or perhaps even paralyzed—limbs using an electronic bypass of the injured spinal cord. Laboratory studies indicate that these brain chips should last for a number of years, but it will take many decades to be sure the chips are truly biocompatible and long lasting.

Assuming the chips prove as enduring as they seem, the prospect exists of "reverse engineering"—meaning that, instead of simply recording brain signals and sending them out for communication or motion, a means might be discovered to bring signals into the brain. This raises speculation of connecting directly to the Internet (vis-a-vis the *Matrix* films), but our incomplete grasp of brain function, cognition, thought, and memory indicates that it will likely take 50 to 100 years before such speculation could even begin to have a solution. The simple bioengineering foundations for this experiment exist today,

A BrainGate chip, shown actual size, interprets the brain's signals and translates them via computer. Someday, BrainGate may allow the disabled to control their limbs through the sole effort of thought.

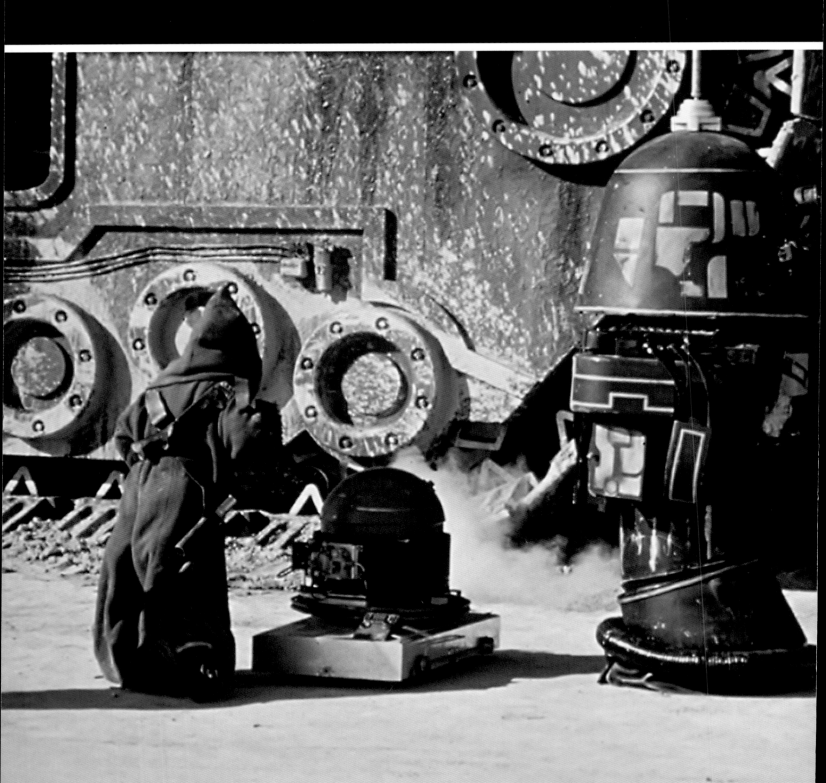

Luke's uncle, Owen Lars, shops for droids to help run his moisture farm on Tatooine. In the *Star Wars* galaxy, robotic help is a simple fact of life.

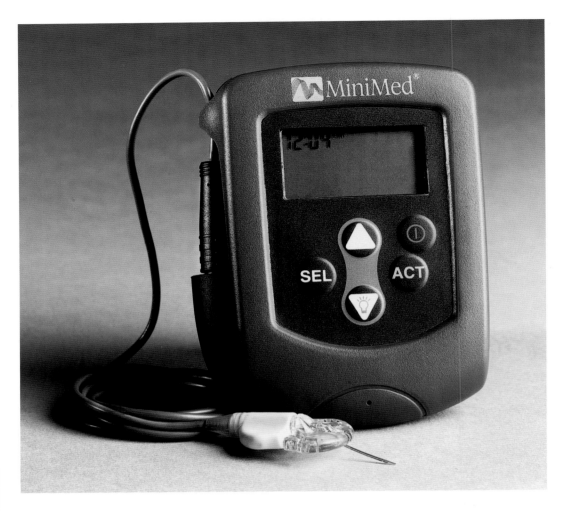

Insulin pumps like the Medtronic MiniMed represent significant medical-technology advances in which man-made devices may replace damaged organs and tissue. Delivering insulin much as a healthy pancreas would, the pump provides a higher quality of life to diabetics than was once possible.

but the complexity of billions of neurons, as opposed to the 100 in BrainGate, remains beyond current understanding.

The next level—beyond simple parts, joints, or chips—involves trying to build an entire organ. For decades, a type of artificial kidney has existed in the form of dialysis machines, and various kinds of pumping systems exist to support or even replace the human heart. However, these all require external parts to provide such things as power, pressure, and fluids. Researchers are working to miniaturize the kidney machine and make it implantable, so it can be hooked up permanently to the blood system to remove the necessary wastes. And the FDA has recently approved the first totally implantable mechanical heart. No long-term studies exist yet, but the future looks promising for a long-lasting, totally implanted independent heart.

Specific organ transplants have been a success story for some time now, with hundreds of long-term survivors. Most donors are human, although some animal organs have also been used. Unfortunately, the problem of the recipient's body rejecting the donated organ remains a clear and present danger. Medications can help prevent rejection, but their toxicity tends to result in unpleasant side effects for the patient, sometimes for a very long time.

A number of institutions seem close to success in the growth of artificial organs. Skin grown from cultured cells currently exists as a commercial product, and synthetic skin—indistinguishable in appearance and feel from real skin—lends credence to the newest generation of medical patient simulators. Disney also uses the synthetic skin to enhance the animatronics in many of its films and theme parks. Growing multilayer organs, however, remains a major challenge. When depth moves beyond one cell layer, the problem lies in getting oxygen and nourishment to the inner cells, while removing toxic wastes—basically, in duplicating the human circulatory system.

At Massachusetts General Hospital and Massachusetts Institute of Technology (MIT), Dr. Jay Vacanti has assembled a multidisciplinary team dedicated to growing an entire organ. This extraordinarily complex process involves a computational mathematician to develop a model of the blood vessel system, which is sent to a stereolithography machine where engineers can "print" it in three dimensions. Biochemists supply special bioresorbable compounds embedded with vascular endothelial growth factor and angiogenesis growth factor for blood vessel growth, involving the work of molecular biologists. This creates a bioresorbable scaffold, which is then placed into a bioreactor—a special nutrient bath for growing cells—with blood vessel stem cells. These attach to the scaffold and form a living blood vessel system. This living system is placed into another bioreactor with the stem cells of an organ, such as a liver. In a few weeks, the cells attach to the blood vessels and a miniature organ develops, kept alive by blood pumped through the blood vessels. The next step involves increasing the organs in size and transplanting them into animals. The initial success of this research indicates there may be no barrier to the eventual growth of artificial human organs, perhaps within a decade or two.

Research efforts in the development of artificial muscle may one day lead to enhancement of the completely mechanical artificial limbs in use today. Fabricated from a number of different materials, the most successful artificial muscles to date are made of electroactive polymers (EAP). Yoseph Bar-Cohen, working in NASA's Jet Propulsion Lab, of Mars Rover fame, is one of many scientists striving to develop synthetic chemical polymers with properties similar to human muscle fibers. Although numerous problems still need solving, particularly in the area of long-term viability, researchers have seen some promising results, which could lead to artificial muscles in 20 to 30 years.

What about building an entire humanoid robot? Here, the future seems to be a rather long time off. However, there will be, and indeed already are, a number of very simplistic, seemingly human or biomimetic systems—that is, mimics of living biology systems, such as humans and animals. Commercial examples include AIBO the robotic dog, with a rather large number of "actions" and "behaviors" programmed to simulate learning—a low level of artificial intelligence derived from rules-based programming. Practical robots for the home, such as Roomba the vacuum-cleaning robot, employ random types of programming to perform a limited and specific task, like cleaning the floor. No one would ever confuse

Bioresorbable: Capable of being absorbed into the human body, such as the synthetic polymer polylactide, a surgical implant that gradually breaks down through hydrolysis into lactic acid.

Latest in Honda's series of mobility projects, begun in 1986, ASIMO is an anthropomorphic robot able to traverse uneven ground and function in a wide range of environments; its humanoid appearance startled the public when it was unveiled in 2000.

ROBOTS AND PEOPLE

these actions with true intelligence, but science is an iterative process, in which we solve one simple problem at a time and then use a combination of simple solutions to solve a slightly more complex problem, in a system that becomes increasingly more complicated and integrated. At present, science resides at the lowest level of developing simple solutions in the process of constructing a humanoid robot, which requires two main elements: the "body," with engineered or grown parts; and the "brain," with computer-controlled programming and intelligence.

In the nonhumanoid arena, a number of new robotic systems are emerging. The military has been using unmanned aerial vehicles (UAV) for more than 5 years, and cruise missiles have been around even longer. The "body" employed here does not mimic a living flying animal, although it is a very complex flying machine. The "brain" that controls the missile or UAV combines sophisticated image acquisition, analysis, recognition, and interpretation systems with very intelligent decision-support systems that take the incoming signals, compare them to the maps and flight controls stored on board, and use these comparisons, minute by minute, to adjust the flight path of the missile or UAV. Some of the new UAVs can actually identify targets, and use this information to release weapons to destroy them. This is not intelligence, but rather a very sophisticated method of pattern matching.

In 2003, the Defense Advanced Research Projects Agency (DARPA), of Internet fame, put out the Grand Challenge—a $1million prize to the first group to build a robot that could drive itself across 150 miles of wild desert from start to finish line, in under six hours. By May of 2004, 25 teams from universities, private companies, groups sponsored by large industries, and even a high school had qualified to build unmanned ground vehicles— UGVs—to run the race. In the end, the most successful team, from Carnegie Mellon University, only managed to travel seven miles before getting stuck in the treacherous terrain. A second Grand Challenge, scheduled for October 2005, will double the prize money. With DARPA increasing the prize each time until a winner claims it, a useful UGV will likely be developed in the not-too-distant future.

Progress has been made in building a complete, full-sized humanoid robot that can walk and navigate difficult terrain, such as climbing up and down stairs. The two best examples, the Sony Dream Robot QRIO and the Honda Humanoid Robot ASIMO, are both completely self-contained and designed specifically to mimic human gait and locomotion. They are clearly robots, made of shining white plastic or brushed aluminum, with hinged legs for walking and arms for waving; their software is functionally programmed for stabilized walking and comprehension of very simple commands. In performance, they obviously can't compare with the full range of human capabilities, but the complexity of their integrated systems remains a significant accomplishment.

Walking, jumping, playing golf, throwing a ball, dancing— Sony's QRIO robot can accomplish all of this and more. A self-contained, autonomous robot, QRIO succeeds in handling extremely complex interactions within its environment.

As these kinds of machines become more human-like, the question immediately arises of whether they can express emotion. Rosalind Picard, director of affective computing research and co-director of the "Things that Think" project at MIT, has written extensively about emotional computing, based on her research in interpreting facial expressions, body language, gestures, and other nonverbal cues associated with specific types of emotions. Her colleague, Rodney Brooks, did groundbreaking work on a robot head called Kismet, which was designed to express emotions. The possibility exists that a mechanical robot programmed to visually show emotion, through facial expressions and gestures, could appear on our horizon in the near future. No one, of course, can add true feelings to such a system; the robot's "emotion" would be nothing more than simple facial expressions in response to certain contextual input.

All of these attempts beg the question of whether robots can be designed with true intelligence, in the human sense. Inventor and computer scientist Ray Kurzweil argues that it is inevitable, and cites his own work as well as that of numerous other researchers, such as Hans Moravec of Carnegie Mellon University's robotics laboratory, who compares estimations of human computational ability (4×10^{19}) with the power of current supercomputers such as Red Storm of Sandia National Labs (3.5×10^{15})—one thousand times slower than the human brain. According to Moore's Law, which (roughly) claims that computers double computational power every 18 months, we're faced with the inevitable conclusion that computers will have more brainpower than humans within the next 20 to 30 years. But does this mean they will be more intelligent? Programming a robot to compute as well as a human differs hugely from designing it to engage in the full realm of human emotional and intellectual intelligence.

A robot that looks and behaves completely human, like the robots we see in the movies, remains many decades if not centuries away. Many partial solutions, where specific functions or appearances give some very limited aspect of human performance to a machine, will come first. We'll see some robots that look real, but do not behave as humans; some very intelligent systems that do not look like robots; and, finally, ever increasing discoveries in implantable devices or "parts," whether synthetic, mechanical, or biological, that will reinforce the notion of some humans becoming cyborgs. In each of these areas, specific examples have achieved a realistic representation of some human part or function; however, progress in the necessary systems integration of most or many of the parts has yet to produce a robot or cyborg that even slightly resembles a human…and certainly not one that resembles Darth Vader. Whether this science will take 20 or 200 years remains uncertain, although all indications point to a distant future when humans will interact with some level of human-like robots. In the meantime, philosophers, theologians, and politicians will doubtless continue to debate the many ethical questions involved in bringing this science to life, even as filmmakers like George Lucas intrigue, delight, and sometimes frighten us by bringing it to life on the screen.

Looming large as the consummate conflicted villain, Darth Vader wields the lightsaber of his Jedi past, now colored Sith red, and the mechanical equipment—helmet, respirator, artificial limbs—that mark his capitulation to the dark side.

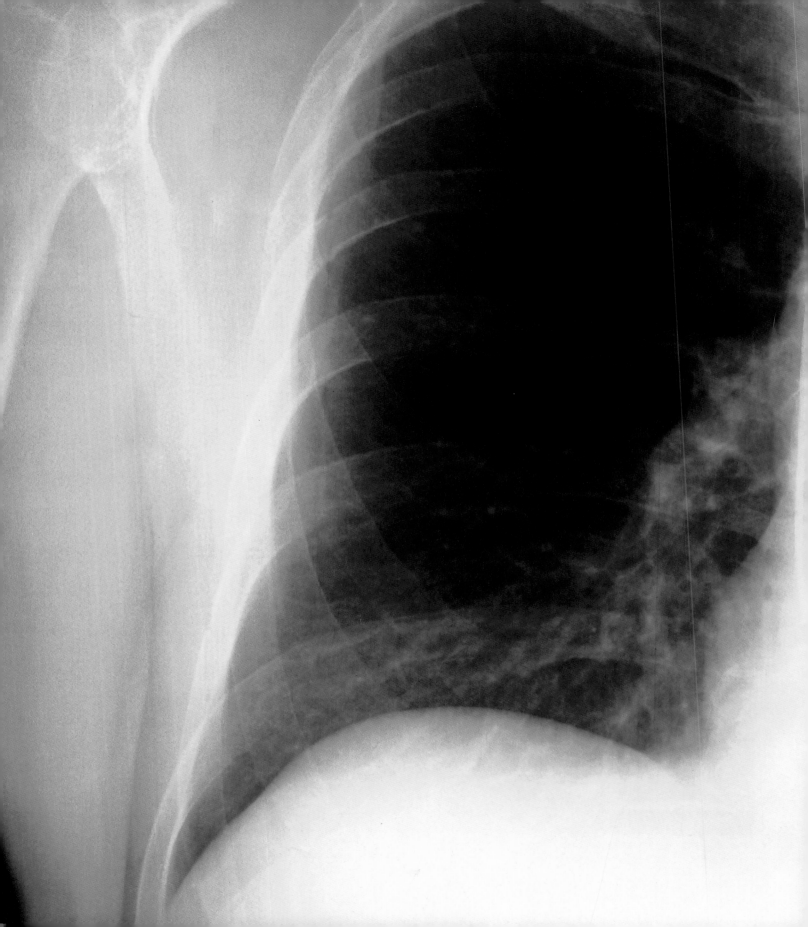

IMPLANTS AND PROTHSTETICS

A GALLERY

IMPLANTABLE PACEMAKER

For the half-million Americans who use them, pacemakers can literally save their lives. When attached directly to the heart, implantable devices like this can monitor heart rhythms for up to ten years on a single battery. When it detects an irregular beat, the pacemaker delivers an electric pulse to the heart muscles, restoring a normal rhythm.

OTTO BOCK GREIFER HAND

Although its looks don't immediately bring to mind a human hand, the Greifer hand has a gripper function that meets the needs of many amputees. Its design allows for precise control and manipulation of objects. A myoelectrically controlled device, the Greifer hand relies on electrical signals from the patient's muscles in order to move.

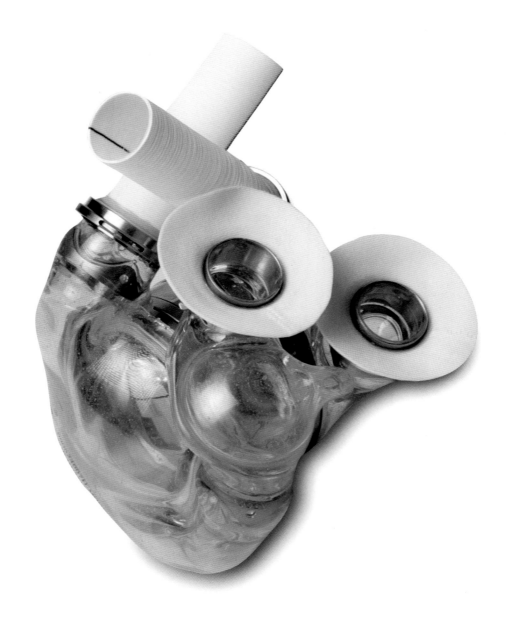

The AbioCor heart aims to be the most fail-safe, long-lasting and durable heart yet built. The heart itself is part of a larger system that includes an inductive battery charger and a computer controller. Unfortunately, these features come at a cost: AbioCor is larger than a typical human heart, and can only be implanted in men with large chest cavities.

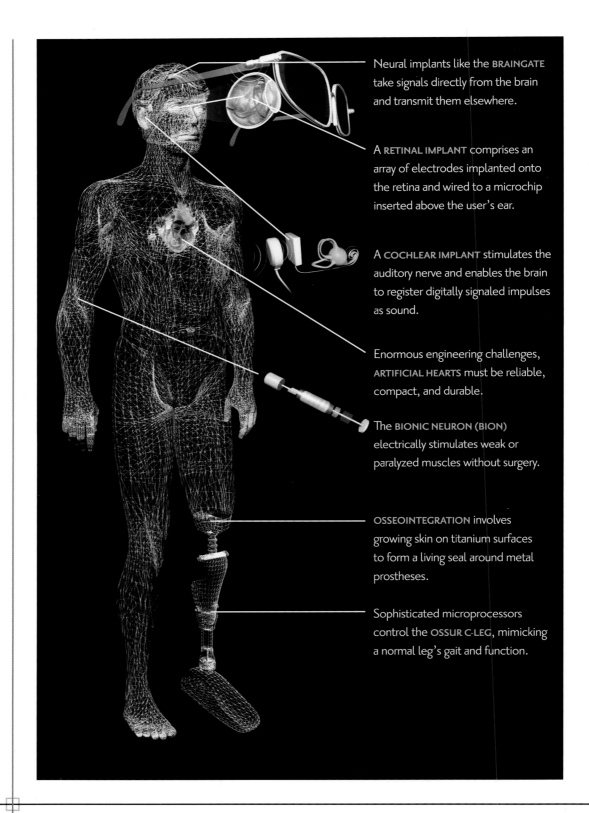

Neural implants like the BRAINGATE take signals directly from the brain and transmit them elsewhere.

A RETINAL IMPLANT comprises an array of electrodes implanted onto the retina and wired to a microchip inserted above the user's ear.

A COCHLEAR IMPLANT stimulates the auditory nerve and enables the brain to register digitally signaled impulses as sound.

Enormous engineering challenges, ARTIFICIAL HEARTS must be reliable, compact, and durable.

The BIONIC NEURON (BION) electrically stimulates weak or paralyzed muscles without surgery.

OSSEOINTEGRATION involves growing skin on titanium surfaces to form a living seal around metal prostheses.

Sophisticated microprocessors control the OSSUR C-LEG, mimicking a normal leg's gait and function.

THE BLEEX (BERKELEY LOWER EXTREMITY EXOSKELETON)

The BLEEX concept consists of a pair of mechanical metal leg braces, a power unit, and a backpack frame that makes a seventy-pound backpack feel as if it weighed only five pounds. Dozens of sensors in the legs constantly calculate how to distribute the weight to minimize strain on the wearer. Although developed for the military, BLEEX has applications for anyone who needs to carry heavy loads—firefighters and relief workers, for example.

HUMAN–ROBOT RELATIONSHIPS

Cynthia Breazeal with Robert Naeye

Cynthia Breazeal, Associate Professor of Media Arts and Sciences, and Director, Media Lab Robotic Life Group, Massachusetts Institute of Technology. Designer of the world's most socially advanced robot, Dr. Breazeal focuses her research on human-robot relations.

I FIRST BECAME INTERESTED IN ROBOTS WHEN I WAS A LITTLE GIRL, ABOUT 10 YEARS OLD. I saw the original *Star Wars* movie and loved the robots, especially R2-D2 and C-3PO. But even then I was savvy enough to know that those kinds of robots didn't really exist, and I didn't know if they would ever exist in my lifetime. I became interested in robotics again when I went to college at the University of California, Santa Barbara. I wanted to be an astronaut, which meant I needed to earn an advanced degree, and I thought space robots would be cool. I applied to top graduate schools and ended up at the Massachusetts Institute of Technology (MIT), with an interest in developing planetary rovers; Rodney Brooks was my advisor. The robots in his lab were insect-like intelligent machines, and I thought that if we were ever to see robots like R2-D2 or C-3PO, it was going to happen in a place like this. Rodney's robots rekindled the *Star Wars* galaxy in my mind.

Unlike HAL from *2001: A Space Odyssey,* who had an eerie, arrogant, disembodied sort of presence, R2-D2 and C-3PO had likeable personalities and risked "life" and limb to help people whom they genuinely cared about. They weren't just machines doing work for humans. They had rich personalities and emotions. They related to people; they had friendships with people. The *Star Wars* robots captivated me and shaped my vision. To me, the genius of R2-D2 is that it enables us to relate to a highly mechanistic droid. It doesn't communicate through language and traditional human social cues, but you know what it's thinking and feeling as it goes through its beeps, clicks, and gyrations. In fact, it demonstrates how socially sophisticated humans are, that we can hear R2-D2's beeps and see its behavior and infer so much about what's going on inside its electronic brain.

I find the relationship between R2-D2 and C-3PO fascinating. I'd expect R2-D2 to be the subordinate droid because it's smaller and more mechanistic. C-3PO, both taller and more human-like, has the appearance of a superior droid, but R2-D2 is actually the leader, the one who basically says, "I have to go save the princess." At the same time, C-3PO is saying, "Oh, why do we have to do that? It's dangerous." The little droid demonstrates loyalty, bravery, and dedication, even if it conveys those qualities in a mechanistic fashion. The rich mental psychology that we can create for that character makes R2-D2 so endearing.

Based on the bickering peasants in Akira Kurosawa's 1958 film, *The Hidden Fortress*, R2-D2 and C-3PO raised the bar for what humans came to believe robots could and should be in the real world.

The Beast, a product of Johns Hopkins' Applied Physics Laboratory, first became operational in 1964. Beast successfully wandered the lab's hallways using sonar navigation.

C-3PO's function, as a protocol droid that can communicate with millions of aliens and know all of their customs, is equally fascinating. We learn in *Return of the Jedi* that C-3PO is also a masterful storyteller, mesmerizing the Ewoks with heroic tales of Luke Skywalker, Princess Leia, and Han Solo. They're both intriguing characters and they both have wonderful emotional arcs, so I can't say that I'm a fan of one over the other. They need each other to bring out their best qualities. If it was just R2-D2 without C-3PO, you wouldn't appreciate them as much as you do when they're together and interacting. They're the Abbott and Costello—or perhaps the Laurel and Hardy—of the *Star Wars* saga.

As time has passed, science fiction robots like Lieutenant Commander Data in *Star Trek: The Next Generation* have exhibited more and more human-like characteristics, so they have become mirrors that we use to look at our own humanity. This raises profound questions about whether machines can ever have emotions, empathy, the ability to relate to people and

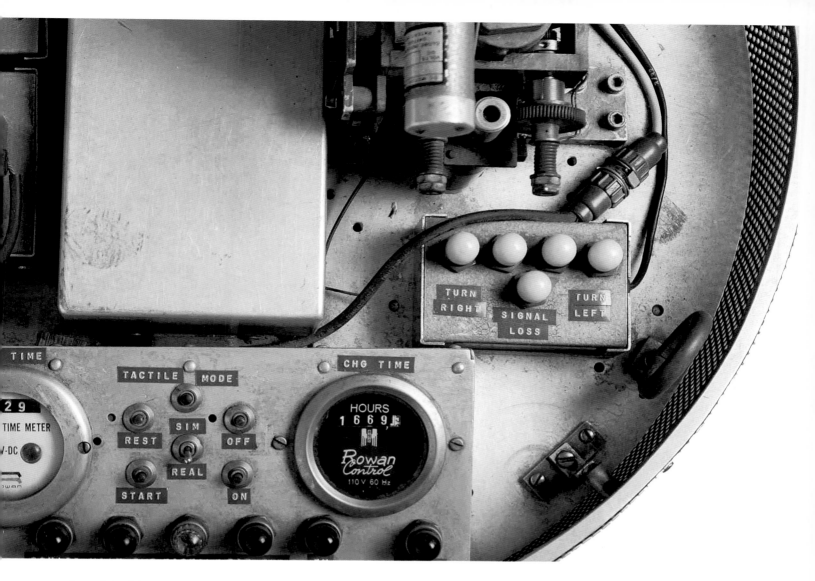

to form friendships. I loved the character of Teddy, a robotic teddy bear in the movie *A.I.* who established genuine friendships with people and other robots. My research at the MIT Robotic Life Group involves trying to make robots like Teddy a reality.

I worked as a scientific consultant on *A.I.,* and during the filming I met four-time Oscar-winner Stan Winston, the preeminent special-effects animatronics expert. Stan created special effects for movies like *Jurassic Park* and *Terminator 2*. I told him that I was researching sociable robots and that we should try to build Teddy for real; he enthusiastically agreed. This is a dream we both share—mine originating from science and invention, and Stan's from storytelling and performance. Stan's studio made a robotic puppet and delivered it to my lab. Stan chose the name "Leonardo" after Leonardo da Vinci, the quintessential Renaissance man who combined sublime art with science, technology, and adventure. Stan and I envisioned Leonardo as a physical robot who would embody our creative endeavor at the intersection of art, science, and invention. We wanted to bring together our different

To recharge, Beast employed a photocell to search for an outlet. As its name implies, Beast had few social skills, but it demonstrates how far robotic technology had come in the 1960s—and how far it yet had to go.

backgrounds in building this robot and capturing the sense of a character like Teddy, a sociable robot that could connect with people in a very human-like way.

Leonardo is influenced by Teddy and also by a previous robot, Kismet, the first sociable robot created to explore interpersonal social interaction with people. Unlike a lot of robots today that still look like machines, Leonardo has a highly organic appearance, thanks to the artistry of Stan Winston's studio. The cosmetically finished Leonardo is a furry, fanciful creature about two-and-a-half-feet tall, with big ears and large, expressive brown eyes. Its 65 degrees of freedom, or points of motion, give it even more range than some of the advanced humanoid robots produced by Honda and Sony. Nearly half of those points of motion are in Leonardo's face, which makes it the most expressive robot in the world today—the Stradivarius of robots.

On the other hand, Leonardo doesn't walk. Our focus has been on interpersonal interaction and particularly on questions of social intelligence. How can this robot cooperate with people as a partner? How can it learn from human instruction? How can it communicate with us in natural terms, and truly understand people as human beings? We're turning Leonardo-the-puppet into Leonardo-the-autonomous-robot by installing sensory and computer systems. The next phase of our collaboration involves getting Leonardo to move and act like a character rather than a machine. Already, Leonardo is becoming much more expressive and capable in terms of how it moves and responds to people.

Machines today still treat people like any other object in the environment. They have no sense that humans are special entities that need to be understood in such psychological terms as belief, intention, emotion, and desire. The fundamental core of social intelligence involves understanding people at that level, so a fair amount of our research examines the scientific questions of how to build a robot with deep social intelligence and social cognition—basically, the ability to understand people in human terms. Dogs and other social animals interact with us in terms of our having minds. How do you build a robot that can interact with people in that fashion? We're searching for answers to deep scientific questions about how to build a robot that has those kinds of capabilities that, right now, only people and social animals possess. The way R2-D2 and C-3PO socially interacted with people is a profound part of what made those robots so popular, and so helpful and valuable to the human characters in the films.

In my own work with human-robot interaction, I recognize the deep significance of social questions. A lot of the work in robotics to date has focused more on cognitive issues such as performing tasks and solving problems. But we will never fully understand humans if we don't understand our social aspects, and using robots as platforms to test theories of social development can help us learn more about ourselves

There is no fixed definition of "robot." In fact, the concept continues to evolve, and it even changes from culture to culture. In the 1970s, a "sophisticated" robot was a manufacturing device that repeated a fixed pattern over and over again, like the robots commonly

Leonardo, currently the most advanced of the sociable robots, has mastered many gestures and facial expressions. Its abilities in interpreting human responses and conveying nonverbal information assist Leonardo in communicating with its handlers, who can accurately track the robot's states of confusion and comprehension, all the while teaching it to perform new, socially appropriate tasks—and improving human-to-robot feedback.

found in automobile assembly lines. Today, we think of those technologies more as machines than as robots. If someone says "robot" to me, I think of something that moves and interacts in our world. The abilities to learn and make intelligent decisions, and then adapt to changing situations, are crucial. In my mind, true "robots" are much more like living creatures than machines; they're not just tools and appliances. The ability to behave intelligently in a dynamic and complex human environment is an important part of being a robot.

A lot of my work is motivated by considering what the consumer robot market might look like twenty years from now. Many robotics experts want to bring robots into our homes to help people, particularly the elderly. But this raises the question of whether elderly people will want to operate robots. Will they be willing to sit behind a computer and program a robot? To overcome this reluctance, we need to find ways in which a robot can become a useful addition to a person's life in the same manner in which a seeing-eye dog can help a blind person. The dog plays an important pragmatic role, but blind people often have deep relationships with their dogs. They enjoy having these dogs around as part of their lives. I'm pursuing the goal of how to build robots with which we can have beneficial relationships, so people see robots as lifelike entities that help us in the manner of capable partners.

To move forward in this area, we first need to understand how people and animals make

"Good 'bot!" Sony's robotic dog, AIBO, recognizes voices and faces, and obeys commands. Owners can switch it from puppy stage for play to mature mode for performing tasks.

decisions. Nature produces entities—people and animals—that are far more sophisticated than robots today. So what can we learn from nature to help make our machines better? How do people and animals learn? How do they interact with others? As we develop better answers to those questions, the insights, theories, and proposed mechanisms will be computed, modeled, and implemented in robots.

For robots to successfully mesh with the human environment, people must be able to understand and predict their behavior. If a robot makes a mistake, we need to be able to understand why it did so. The more the robot's behavior and the reasons for that behavior mirror natural analogs, the more we will be able to understand these machines, and the better we can communicate with them, trust them, and work with them. Further, the robot must openly express what's going on inside its head so that the human understands its thought process.

> We must appreciate that a robot relationship with a human may never be like a relationship between humans, or even between humans and animals.

Developmental psychology—particularly the social development of infants—is a major source of inspiration in my research. I often think of the many ways that science fiction has portrayed robots as a sort of mirror of our own humanity, like Lieutenant Commander Data's quest to become more human. This gives me a deeper appreciation of how amazing people truly are. Some members of the general public have the impression that "arrogant scientists think they can do something better than nature." But from my own experience building robots, I can say that there's nothing more humbling than trying to capture even the tiniest instant of human behavior in a machine. I recently had a baby, Ryan, who is now fourteen months old. Watching my own baby grow and develop every day is one of the most profound experiences I have ever had. It has given me deeper insights into what is so special about people, making it all the more imperative to capture those qualities in machines so they can fit into our culture.

Based on my own experiences with Ryan, it's clear that the earliest interactions are social. It all starts with face-to-face exchanges of expressions and behaviors. For me, these interactions highlight the importance of the social nature of our development and learning. Just as infants like Ryan can respond to human facial expressions, our lab has had some positive results with robots making similar responses. Still, we're a long way from building robots that can perceive and respond to human facial expressions with the subtlety and sophistication of humans in face-to-face conversation.

It will be many years before we'll see machines that have the abilities of a two-year-old child. You can build a machine that's highly dexterous in a very specific way, but no one knows how to build one that's as broadly dexterous as a two-year-old human. We can design

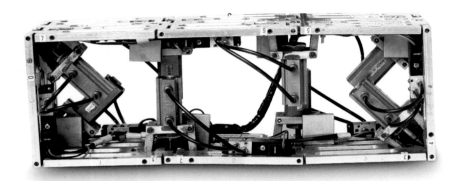

machines to traverse long distances, but they can't scramble over the kind of terrain that a two-year-old can. We have computers that can learn, but nowhere near as fast as a two-year-old. And children are much more sophisticated socially. A two-year-old girl knows how to communicate with you, how to understand when you're happy or angry with her, and how to get her way. Children also have very rich perception: They can see, hear, touch, taste, and smell. Robots by comparison are incredibly impoverished in these areas.

As I watch Ryan, I am fascinated by the nature of preverbal infant thought, partially because it raises questions concerning how we can make robots learn in a more organic way. People expect a machine to learn a specific task very quickly and efficiently. But look at the way infants learn—it's messy, creative, and serendipitous. We need to make machines that can learn in a much more open-ended fashion. We need to instill in robots the human aspect of sophisticated exploration and a sense of creativity, of having insight into what to try next. I'm sometimes amazed at how insightful my fourteen-month-old is when he's trying something new.

I remember watching Ryan learn how to feed himself Cheerios when he was around seven months old, and how I played a helping role. First he had to learn that the pincer grip—picking up a Cheerio between his thumb and forefinger—was the most effective way to put a Cheerio into his mouth. He rarely succeeded using his fist. To help him learn the pincer grip, I would "deal" out one Cheerio at a time, because he was more likely to try to grab a bunch in his fist if he saw the cereal in a pile. Sometimes I would hand him a Cheerio in such a way that he could easily grasp it with his thumb and forefinger, allowing him to

Rectiblob uses a novel approach to get around: Its entire body provides its mobility. It can roll like a wheel or a caterpillar tread, or stand up and fall forward to get over obstacles.

ROBOTS AND PEOPLE

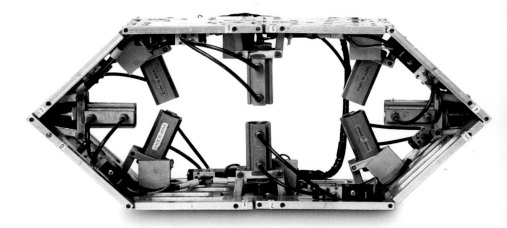

experience early successes using the pincer grip. Then he figured out how to put the two together in a totally novel context. In one particularly insightful moment, he grasped a Cheerio in the fist of his right hand, used a pincer grip to transfer it to his left hand, and then continued to use the desired grip to delicately put the Cheerio into his mouth. "Genius!" I thought. What would it take to build a robot with this kind of creative and constructive problem-solving ability? Because Ryan hasn't developed language yet, he can't tell me the nature of his thought process, but he's clearly thinking and trying to understand.

A lot of recent robotics research, including my own, involves having machines learn from observation and imitation, just as infants do. Machines can easily learn physical motor skills. We can already program a robot to perform a fixed motion in order to paint a car just by physically guiding it through the motion and having the robot record it—basically "showing" it how it's done in a rudimentary way. We're getting more sophisticated in research labs, teaching robots how to begin learning new motor skills from visual observation, such as hitting a tennis ball with a forearm stroke. But the goal of my work is to have robots go to a deeper, psychological level. I want to see robots develop what developmental psychologists call a "theory of mind": the ability to understand humans in terms of their underlying psychological states. If you're watching someone else in a particular situation, you can use your own decision-making process to think, "If I were doing that action in that situation, this is what I would be trying to do." That's a sort of empathetic understanding, the ability to stand in the other person's shoes in order to fathom what their motivations and goals might be.

Within the complex machinations of the droid factory on Geonosis, anyone could lose their head—as C-3PO famously does. Such scenes reflect the

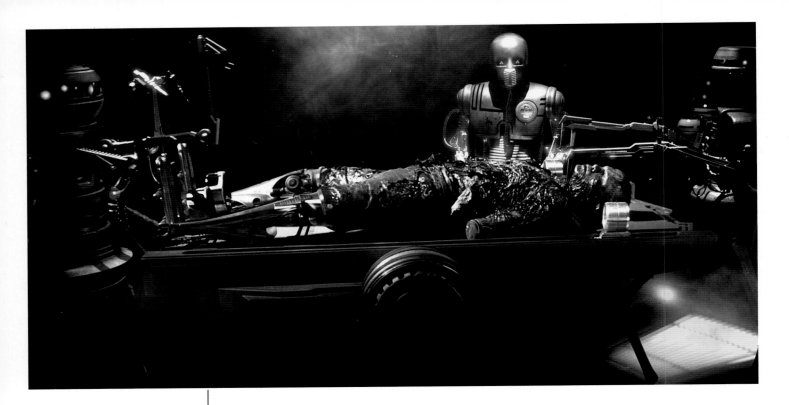

We must appreciate that a robot relationship with a human may never be like a relationship between humans, or even between humans and animals. But relationships with robots could have certain advantages. After all, some of our most rewarding relationships are with non-human entities, like dogs and cats. Relationships with pets don't have a lot of social baggage, such as losing face, being embarrassed, feeling judged, or feeling vulnerable because of age or illness. In their dealings with robots, people may be able to take some comfort in avoiding these unpleasant aspects of human social interactions. Robots may be able to combine certain positive aspects of human-human relationships with human-pet relationships to create a novel kind of human-robot relationship. My work looks at how robots can enhance the palette of the very special connections that we already have. I'm not saying that robots should ever replace those relationships, but rather provide additional benefits, ones that other relationships may lack.

Human-animal interactions provide some guidance. Hospitals and nursing homes often bring in animals to provide therapeutic benefits to patients. They can help children undergoing cancer treatments or elderly patients with dementia. Animals provide social support, and mounting evidence indicates that they help lower blood pressure and produce a calming effect. People with a relaxed, positive mind-set heal faster and are less prone to depression. Takanori Shibata in Japan has built a robot that looks like a baby harp seal, which he hopes to commercialize and bring into the homes of elderly people so they can interact with it. The seal also serves the same social-facilitator function as bringing a pet into an environment, inspiring people to converse with each other by talking about the animal. But animals are not allowed in many places, which highlights another benefit of introducing robot

Medical robots work on the nearly lifeless Anakin, after he is overwhelmed by Obi-Wan and the fires of Mustafar. The genius of technology replaces his severed limbs, charred skin, and damaged organs to create the machine-man, Darth Vader.

Japan's prime minister, Junichiro Koizumi, greets ASIMO, Honda's humanoid robot, during a visit to the Czech Republic. ASIMO capably demonstrates the robotic potential to integrate within human society.

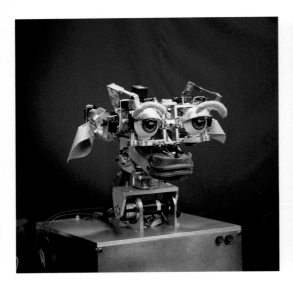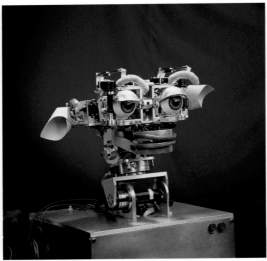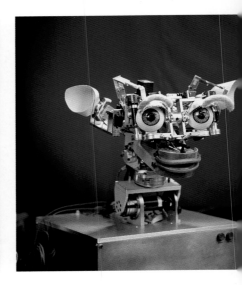

surrogates. And in addition to interacting socially with patients, robots can carry data-recording and monitoring devices, which could feed information to nurses and help them do their jobs better.

Building socially intelligent machines is an extraordinarily difficult undertaking. Social intelligence isn't just about communicating and being friendly with someone; it's about the many deep ways we interact with and understand the world and its people. For example, how do you work with another person as a teammate? How do you coordinate your actions and mental states to carry out a collaborative activity as a capable partner, rather than being used as a tool by another person? My research group focuses on those fundamental questions.

The ability to learn from others has profound significance in the development of collaborative skills. Researchers are making considerable progress in enabling machines to learn by themselves, an important skill for robots working on Mars, for example. But if robots are introduced into the human environment, people will want to teach them how to do new things. Robots will need the ability to adapt their behavior to a human's preferences. We're trying to build robots that can learn from people who are not necessarily experts in robotics or machine learning.

Humans are not entirely rational creatures like the character Spock in *Star Trek*. Emotions clearly played a vital role in our evolution. People whose emotional systems are impaired do not behave intelligently. Human intelligence is as deeply intertwined with emotional reactions as with cognitive abilities, and scientists have a great deal to learn about the role of emotion and how it interacts with cognition in order to build more intelligent, capable machines.

A robot that people can interact with in an intuitive and enjoyable manner has to be predictable and compatible to our social, emotional, and cognitive processes. Robot owners will constantly assess: What is the quality of my experience with this machine? Am I enjoying its presence, or is it annoying the heck out of me? Those kinds of issues will constrain how we design the behaviors and the communication abilities of these machines, and

The many faces of Kismet shed light on the future of our interaction with robots. Human expression, far richer than what words alone convey, encompasses a significant robotic challenge: imparting a fraction of the emotional scope humans can display with a simple raise of an eyebrow.

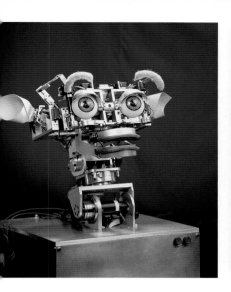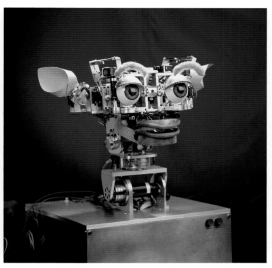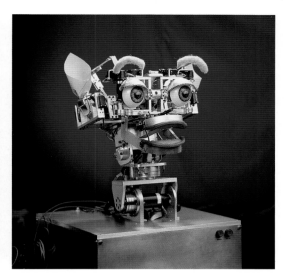

the degree to which we're successful will play a very important role in whether or not people will accept robots into their daily lives. Fortunately, the positive impact of R2-D2 and C-3PO has raised public awareness in America of the possibilities of intelligent and sociable robots.

But we will have to make technological breakthroughs in many areas to realize my ultimate dream of robots that can forge meaningful relationships with humans. The areas needing improvement are not limited to computers and other devices that will make robots able to process information in a more sophisticated manner. For example, the materials used to make today's machines—mostly metals and other inorganic materials—are inferior in many ways to those in plants and animals. When people and animals bump into things and get injured, they can heal. But robots don't heal on their own. How do we build machines that are self-healing? Forty or fifty years from now, machines may actually be based on the same materials as living organisms.

Living creatures have another crucial advantage over machines: longer-lasting power sources. A major breakthrough will occur when machines can fuel or recharge themselves for long periods, or use other renewable sources of power, without being tethered to electrical outlets. Research on fuel cells will definitely benefit the field of robotics.

Natural language understanding and auditory processing are particularly difficult undertakings. Imagine taking a robot into a crowded mall just before Christmas. Humans can hear and understand multiple conversations and extract a single conversation in a noisy environment, but that's an incredibly difficult task for a machine because it can't single out a stream of auditory information—the so-called cocktail-party effect. Although we are making progress, you could not take a robot into a noisy mall right now and have a normal conversation with it unless you placed a microphone right next to your mouth. Even then, having the robot understand the meaning of what you are saying presents a challenge that researchers have been working on for many years.

Better robotic sensors are also important. Human eyes perform dramatically better than

Auditory Processing: The ability of the nerves of the inner ear to transmit impulses to the brain concerning hearing and balance.

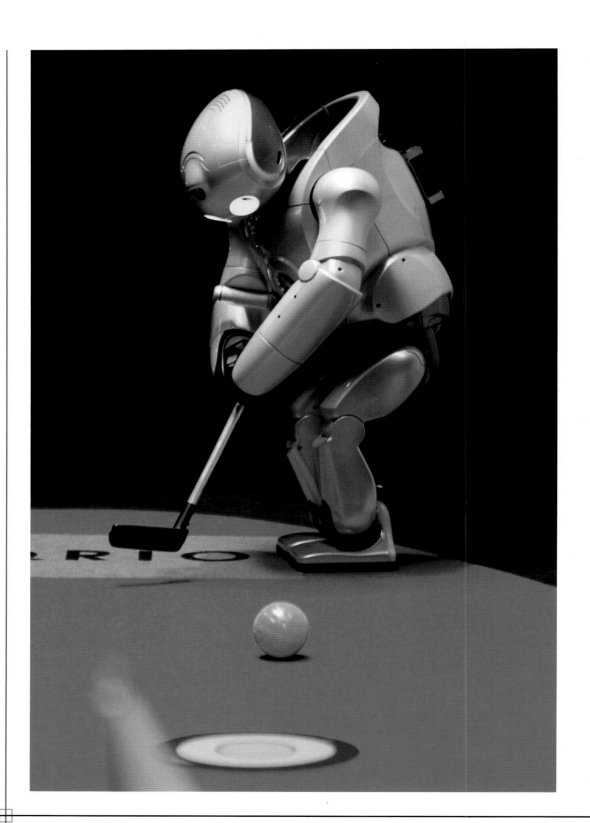

Sony Corporation showcases their humanoid robot QRIO playing golf during a demonstration at company headquarters in Japan. Teaching a robot to improve its swing is only one of the many challenges that engineers must conquer in the future of robotics. More importantly, can we teach robots to interact with humans—even to act like humans?

cameras at focusing quickly and seeing in various lighting conditions. If you take a traditional camera from your living room to your backyard at high noon, the change in light creates a vastly different visual signal, and that same problem exists for robots as well. Vision algorithms can perceive human faces to some extent, but they tend to be sensitive to lighting conditions. Scientists are developing new vision-based technologies, such as integrating the light detectors used in digital cameras and camcorders with infrared sensors, so robots can have more effective depth perception.

A great deal of work remains to be done on the computational side. What will robots do with all the information that bombards them from all directions? The world is a very complex place compared to a laboratory. Think about a chaotic and dynamic home with kids and pets running around. Dealing with the complexities of daily life will be a major challenge, even more so for robots than for us!

Finally, significant advances need to be made on the cognitive side. How smart will these machines be? That will depend on computer power as well as ideas for what kinds of processing those computers will do. With networking technologies, the future robot's brain will be more than a set of processors mounted on the machine. The robot will utilize all the computational tools in its environment. So if current trends continue to the point that computation devices are all around us, robots will be able to tap into those resources to operate more intelligently. Researchers are trying to figure out how to design the architectures, algorithms, and internal computational mechanisms for these machines.

Huge improvements will have to be made in machines' ability to perceive human movement and interact with people either one-on-one or in groups. Distinguishing people as individuals will be a challenge because machines do not yet understand people in terms of shared history, experiences, and relationships. Human intelligence probably evolved the way it did primarily to handle complex group dynamics and to solve social problems.

The field of social robotics is still in its infancy. The deeper scientific questions concern how to make robots that are not only socially interactive, but socially intelligent as well. I look to theories that try to explain the social intelligence of humans and animals as I strive to tackle several "grand challenge" problems of social robotics. How can robots work collaboratively as peers alongside humans? How can robots engage in rich forms of social learning, such as imitation and tutelage, in order to learn new skills and acquire knowledge? How can they participate in beneficial, long-term relationships with people, in which they understand people as people, in order to appreciate our motives, beliefs, feelings, and desires? Given that robots have been exploring other planets for years, it's ironic to consider that our homes just may be the "final frontier" for robots. The complexity of human society raises many new challenges for them, but with every advance in these "grand challenge" areas, R2-D2 and C-3PO are coming closer to a home near you.

THE SEGWAY HT

DEAN KAMEN

DEAN KAMEN President, DEKA Research and Development Corporation. Dean Kamen is an inventor, physicist, and entrepreneur who holds more than 150 U.S. and foreign patents, many of them for innovative medical devices.

WHAT APPEALS MOST TO ME ABOUT THIS BOOK IS EMBODIED IN ITS TITLE, "Where Science Meets Imagination." I believe that we as a society are not doing enough to prepare ourselves and our children to act as informed citizens in our technologically advanced world. In the *Star Wars* saga, Anakin Skywalker built C-3PO as a nine-year-old. How many of us can take apart a computer and identify major components, let alone replace a broken one? We make such little progress in so many technology debates—nuclear power, cloning, disposable plastics—because we aren't well informed, not just about content but also about process. Not everyone needs to be an engineer, but we should all be able to separate fact from nonsense, to appreciate how new inventions come about, and to understand the trade-offs in a design process that takes place in a world governed, not by man-made laws, but by the laws of thermodynamics and conservation of energy. That's where science really meets imagination!

People always ask when we started working on Segway. I can say with great confidence—I don't know. I have to trace the process back to the long fascination I've had with gyroscopes, balance, mobility, and transportation. In college 30 years ago, I used to ride a unicycle. It amazed me that humans could acquire such an unnatural skill, and it fueled my interest in the physics of balance. Later I went into business building small, high-precision medical devices, but I was also interested in helicopters, which are all about stability and balance. And last, but certainly not least, I would see somebody in a wheelchair and think, "I can't believe how little these devices have changed." The wheelchair was invented in the days of Benjamin Franklin, and it astounded me that we hadn't applied new technology to the concept of the wheelchair in almost three hundred years.

So there I was, with diverse interests in physics and transportation, and the result of all that was the iBot—a dynamically stabilized device that can climb curbs, or stairs. It can be raised or lowered to put the driver at eye level with a person who's standing or sitting, and it helps equalize the people who need it with the people who don't. Was that the end of the story? Of course not. One invention usually leads to another, and iBot was no exception.

We had this balancing technology, and we realized that its applications could extend beyond the needs of the disabled; it could provide anyone with many of the advantages we look for in transportation technologies. Cars, subways, and buses all run on wheels and enable us to move around without a lot of effort. Even human-powered devices like bicycles are enormously more efficient than a human on foot. Still, a bicycle, as beautiful as it is, was never intended to function in a dense pedestrian environment. Consider the fantasy world of the

Dean Kamen has long been fascinated by self-stabilizing devices. The Segway traces its lineage back to his efforts to develop a dynamically stabilized wheelchair, the iBot; this device's potential inspired Kamen to design a similar device for the general public.

recent *Star Wars* movies: The droids coexisting with humans often have two legs or balance on a single set of wheels, and occupy the same amount of real estate we do.

I began to contemplate a version of iBot that a person could stand on. Even able-bodied people could increase their ability to get around. It would be like magic sneakers—and it would compete not with cars, but with feet. I found this very exciting, because transportation technologies have changed the way we live and work. Every major development in transportation has changed not just the scope of the world but also the way we perceive it.

Sailing ships gave us a new continent; air travel put the whole world within our grasp. Compare the Model T to the Ferrari, and the Wright Brothers to the 747. These advances are amazing, but how do we usually get around? On our feet, about 90 percent of the time. And how have we applied technology to this method of transportation? Ancient Greeks slapped on sandals and cruised around Athens at two miles per hour. Fast-forward to today. You slap on those Nikes…and you move around at two miles per hour. In thousands of years, nothing has changed for pedestrians in terms of speed.

What did change is the size and density of cities. Look at the microcosm of an ancient city: It was small because people wanted it to be dense. They wanted the ability to move quickly from place to place. We've done nothing to make a significant difference in how people get around in cities; if anything, we go slower now because we've spread things out, and because modern traffic involves cars, trucks, buses—all pretty much creeping along at about eight miles per hour. When I look at what has to be done to a city to accommodate these huge machines—designed to travel at 60 miles per hour—it boggles my mind.

Walk a mile in London or New York City, and you will move more slowly than you would have done a hundred years ago, largely due to traffic jams. Even public transportation can't handle the crowds nearly as efficiently as would be the case if pedestrians could travel just twice walking speed—increasing their speed to four to six miles per hour. What if we could put

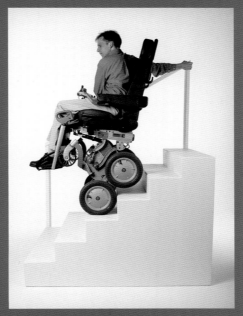

people on the same footprint as the guy wearing the sneakers? What if, within an existing infra-structure, people could cruise along at four miles per hour, or six or eight? Those speeds are slow enough to optimize control, but fast enough to represent up to a 400 percent increase. The potential of the idea becomes intoxicating.

All our different modes of urban transport—cars, taxis, mass transit—end up delivering about eight miles per hour, a very energy-intense, environmentally taxing eight miles per hour. What if we designed a new transportation device that would be environmentally friendly and fun to use? Here's an essential lesson of new technologies: When we invented the iBot for a very specific purpose, we saw another, unexpected place where this technology could make things better. We applied the core iBot technology to the one piece of transportation—walk-ing—that hasn't seen a big idea since the sandal. We asked ourselves how we could make it an attractive alternative to these other modes that are so bad for the environment and the global economy.

For decades, people have left the city to escape congestion, moving out and creating sub-urban sprawl—and spending an hour or more commuting. The suburb is not pedestrian friendly; people still need cars to go from one place to another. We have eviscerated cities whose fun-damental, original goal was to provide a highly dense, rich environment in which people could interact. As a transportation device, cars run contrary to this goal: You're simply not part of the community you're moving through when you're sitting in a car.

Segway lets you function as a pedestrian, so it fits within an existing pedestrian environment. In fact, it can make that environment more attractive by offering an enjoyable option for navigating through it. More areas can be zoned pedestrian-only, and downtowns can compete with shopping malls. Segway has the potential to make pedestrian environments friendly, clean, and efficient again—assuming, of course, that people accept it as a viable alternative. Unfortunately, the mer-its of a particular technology or invention don't necessarily insure that it will be adopted quickly.

To someone in a normal wheel-chair, a staircase might as well be a mountain. Dean Kamen combined his fascination with balance and his engineering skills to create iBot, a self-stabi-lizing, four-wheeled device that not only climbs stairs with amaz-ing grace, but also serves as a kind of social elevator, enabling the wheelchair-bound to achieve eye-level status with a standing person.

The technologies that DEKA spent years developing for iBot and Segway no longer seem exotic; DEKA engineer Benge Ambrogi and his teenage son, Nick, built this two-wheeled device using off-the-shelf parts at a cost of less than $500.

We're witnessing an unprecedented global move toward urbanization. In Asia alone, 800 million people will move into cities in the next decade. If they all choose the American model of the last century, and 800 million people decide they want cars, life as we know it is toast. There's not enough oil; there's not enough steel; there's not enough air. It's a nutty model, but we have nothing in the wings to replace it unless we rethink the whole system. People want new inventions to fit the existing model. This was true when cars were first invented: We tried to make them operate in ways that accommodated the world of the horse. Only when it became clear that they would supplant the horse did people's thinking change, allowing a new technology to take hold and thrive.

It's hard to get people to stop doing something comfortable but bad, simply by telling them it's bad. You get people to stop doing bad things by giving them better alternatives. That's the role of inventions. We should consider technologies today and say, "Look at all these great things. But, hmm...look at the unintended consequences. It's time for an invention!" We need to keep what's good, get rid of what's bad, and move forward. I find it unnerving how few people engage in the debate at that level.

On the one hand, people in love with a particular technology don't want to recognize the

ROBOTS AND PEOPLE

unintended, harmful consequences; they're in denial, or they say, "Not my problem. I'm a technology guy." And then there's the other side, people who say, "We don't want any of that technology. It's bad, period." But none of those people want to live in a cave and rub sticks together to get heat and light. How can we insure that if somebody's going to have a highly emotional opinion, it's at least grounded in fact, especially when technology is changing so fast that the facts are changing fast too? The answer lies in getting more and more kids—future voters and policy makers—comfortable with what technology can do and with the compromises that it inevitably requires.

What is technology? It's simple: Technology is anything that wasn't available when you were a kid. For my grandparents, technology was the airplane. For my parents, it was the television. For us, it's the computer, and the Internet. But kids today view all that as infrastructure. They use it every day and take it for granted, but since it's harder than ever to catch up with what's making it all work, they're less capable of understanding its implications—implications that are more profound than was the case for simpler technologies in a simpler time. This generation actually seems less engaged in understanding the technologies that loom so large in their lives.

America is a democracy. I'd never give it up for anything else; I love it, but in America, everybody gets a vote whether they're informed or not. As a result, our public policy will shift in the direction of what 51 percent of the people think is right. If more than half the population cannot make informed, intelligent decisions about when and how to apply technology, we will be living in a world where we're not making good choices. Technology isn't just about use, it's about understanding—understanding how to be technologically responsible, how to meet our obligations to the environment and to our neighbors. I believe that, as the latest technology screams forward at an ever accelerating pace, we need to find ways to encourage more and more kids to put some time and effort into understanding it.

Where does this leave Segway? It's not a comfortable fit with the automobile-pedestrian dichotomy. It was explicitly designed to coexist with pedestrians, but many find it difficult to link with people. They look around for a model to compare it with—it's sort of like a bike, but it's not. It's sort of like a motorized wheelchair, or a scooter—but not. It doesn't fit the current paradigm, so it's encountered a lot of resistance. Will it ever be the mass-market transportation device I hoped it would? I don't know. One thing I can tell you, though: It has already morphed into something I never anticipated.

At universities all over this country you will find Segways, not with people riding on them, but with robots. Many roboticists are interested in designing robots that can work alongside humans, which means they need to function in a world of obstacles and tight spaces. Enter Segway. Who knows? Segway may wind up being the mobility option of choice for the robots of the 21st century. Technology is a funny, unpredictable thing. Nothing makes me happier than imagining what the future could look like, and then trying to make it happen. I recommend it to everybody.

BRINGING A MOVIE ROBOT TO LIFE

Grant Imahara

Grant Imahara, Animatronics Engineer and Model Maker, Industrial Light & Magic. Grant Imahara worked in animatronics and model making on all three of the *Star Wars* prequels and was one of the few official operators of R2-D2.

THE *Star Wars* GALAXY IS FULL OF WALKING, TALKING DROIDS THAT ARE CAPABLE OF conveying emotion and intelligence. They're part of everyday life, and, as characters, they must hold their own against the flesh-and-blood actors with whom they share scenes. In reality, however, the droids you see on screen aren't really robots at all. Often, what looks like a thinking, speaking, animated robot in a movie is in fact a puppet being controlled by many operators just out of view of the camera's lens. In addition to these radio-controlled and traditional rod or cable puppets, we also use high-tech computer graphics and animation. Sometimes, we even use an actor in a suit.

The first step in bringing a movie robot to life lies in interpreting the vision of the filmmaker. The director collaborates with a concept artist, who will generate the initial artwork for a character, which typically undergoes numerous revisions. Once the artwork is finalized, the movement and performance requirements must be defined based on the character's role in the film. Next, the production designer, the visual effects supervisor, the producer, and the director decide which approach to take: an animatronic (radio-controlled) puppet, a traditional rod or cable puppet, a costume, or a digital rendition. The final artwork then gets passed to a prop maker, model maker, fabricator, or digital artist, and construction begins. In some cases, a computer-aided design draftsperson will convert two-dimensional conceptual artwork into plans for a three-dimensional character. Depending on a project's complexity, the process from concept to completion can take anywhere from several weeks to several months, and utilize the skills of numerous artists and engineers.

ANIMATRONIC PUPPETS

The high mobility of animatronic puppets offers the filmmaker numerous shooting options, as these puppets come without the baggage of control cables or wires. Because they are completely self-contained, they can perform with actors on the set. Once the choice has been made to create an animatronic puppet, artwork for the character goes to the fabrication team. While sculptors, mold and model makers, and painters focus on the outside of the puppet, mechanical and animatronics engineers create and assemble the complex mechanisms inside the puppet that allow it to move and speak. A great deal of planning goes into

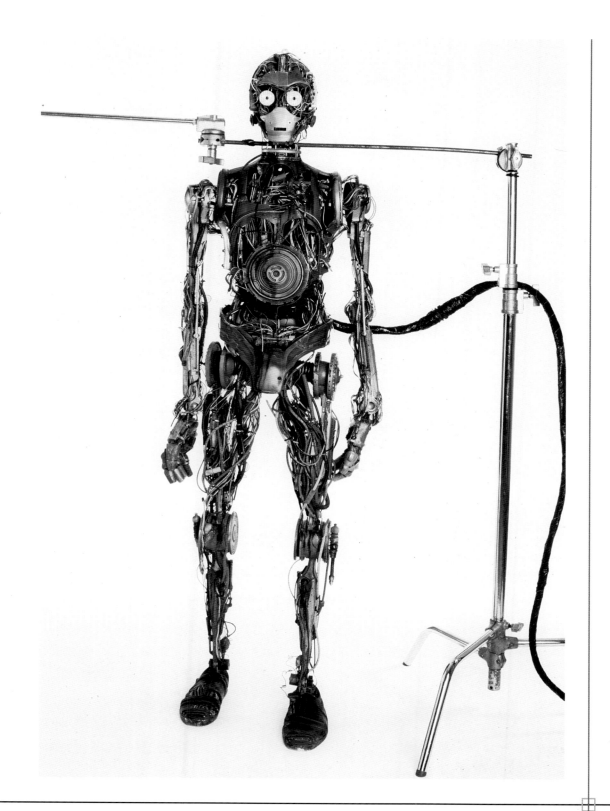

In *The Phantom Menace*, Japanese-style puppetry called Bunraku portrayed the corporeal C-3PO. Though the heavy marionette proved unwieldy and difficult to use, it gave the droid a unique lifelike quality, and intricate internal details impossible to achieve with a live actor.

Technological advancements made in the world of cinema enabled production and ILM technicians to upgrade R2-D2's radio-controlled abilities from film to film; digital animation even allowed the little droid to fly and climb stairs in the prequel triology.

ROBOTS AND PEOPLE

the building process, and every animatronic puppet must be designed with repeatability, reliability, mobility, and control features in mind.

A movie puppet's performance hinges on repeatability. Once the director decides on a specific action for the character, the operator has to be able to perform that action, however complicated, with perfect repeatability every time. Actors rely on these timing cues to do their jobs. When the puppeteer repeats the control moves, the puppet must always respond consistently to the input, which is the responsibility of the mechanical and animatronics engineers.

Reliability is a special concern with complex puppets. An inopportune breakdown can ruin the momentum of a scene—or even an entire day—which translates into cost overruns for both the schedule and the budget. Through the years, we have refined and upgraded the R2 units to the point where breakdowns happen only rarely, but not every movie robot has the benefit of such longevity. Most production budgets don't allow for rigorous research or testing, so it's up to the skill and experience of the fabrication team to build a reliable, robust system within a short time frame.

Though mobility is a key advantage of animatronic puppets, it also represents a critical design concern. Take R2-D2, for instance. In order to get around, the radio-controlled R2 is equipped with hidden wheelchair motors, one in each foot pod. Quiet and efficient, these battery-powered motors help extend the length of unbroken shooting time. The movie droid must also adapt to different sets or locations, as with R2-D2 in *Star Wars:* Episode I *The Phantom Menace.* The R2 units originally sported smooth wheels, but they had to be retrofitted with knobby tires for filming scenes in the sand. Yet these knobby tires worked only on sand; smoother surfaces caused them to vibrate, making the metal heads of the R2 units rattle when they moved.

Finally, control must be maintained at all times to insure the safety of everyone on the set. The radio interference of outside signals sometimes provokes unexpected behavior, and large animatronic puppets running out of control can cause huge safety problems for the actors. In the past, even R2-D2 experienced problems with interference. Renegade R2 units had a habit of bumping into actors, and were even known to run over the director's feet on occasion. For this reason, animatronics engineers now program the radio transmitters with failsafes, defining safe conditions for the puppet when the radio control signal gets lost due to interference or power failure.

ROD AND CABLE PUPPETS

Just as animatronic puppets have their place in the movies, so do traditional rod and cable puppets. Rod puppets have steel rods attached to key joints or control points, which are held and operated by the puppeteer. Cable-controlled puppets have long, steel-sheathed cables, similar to the brake control cables on a bicycle, which lead to big mechanical levers or hand controls. Because these traditional-style puppets are manipulated directly, as opposed to remotely, they offer greater subtlety in performance. Their drawback lies in mobility, since

Failsafes: Features specifically designed to have the capabilities of automatically counteracting the effect of an anticipated potential source of failure.

the puppeteer has to be close to the puppet. Special sets may be required to hide the puppeteer, and green or blue backgrounds may be used to facilitate their removal from the scene later on.

Star Wars: Episode I *The Phantom Menace* features a C-3PO portrayed by a traditional rod puppet. In the other movies of the *Star Wars* saga, this character is played by Anthony Daniels, an actor in a droid suit. *The Phantom Menace*, however, called for a less-than-finished appearance for the character, including a see-through torso, which precluded the use of an actor. In order to get around this problem, we borrowed an ancient form of Japanese puppetry, called Bunraku, in which the limbs of the character directly connect to those of the puppeteer. Dressed in green or black to fade into the background, the puppeteer and his or her assistants manipulate the puppet from behind. C-3PO's open head and torso contained visible, spinning gyroscopes that made the puppet quite heavy, so that walking was possible for short distances only.

ROBOT COSTUMES

Sometimes it's easier and more straightforward to put an actor in a suit. Because the actor directly affects the movement of the character, more nuanced performances are possible than with puppets. The drawback here lies in the costume itself: Typically hot and heavy, these costumes can be supported only for a limited time before the actor needs to take a break.

Seen from the back, the nascent C-3PO from *The Phantom Menace* is a complex of wires and panels—a woefully exposed condition for a protocol droid.

Difficulties increase when the actor's head is fully enclosed, as with the C-3PO costume, which provided only two small eyeholes to see through and a mouth slot to breathe from.

While putting an actor in a robot costume may simplify construction and performance issues, the costume itself can still end up being quite complex. For example, C-3PO is composed of 17 parts—not including the battery pack and fastening hardware—that must be worn by the actor. Most of the costume consists of fiberglass and metal, although vinyl is used for soft parts that require flexing in many directions, such as hands and feet. The midsection has a rubber girdle with bits of wire and flexible tubing sewn in place.

R2-D2's costume sometimes requires a small actor—Kenny Baker—to fit into R2's small

For more than 25 years, Anthony Daniels has donned a cumbersome golden suit and transformed himself into an icon—a protocol droid known the world over as C-3PO, or Threepio for short.

canister. The robot's dome-shaped head swivels easily on a ring bearing; the actor rotates it manually, and also manipulates the projector eye with a lever from behind. Hollow foot pods allow the actor to slip his feet inside them. Technically bipedal, like C-3PO, the two-legged R2 unit has limited walking capabilities, so self-balancing isn't an issue. However, the actor can make R2 rock from side to side—something even the animatronic versions can't replicate. This signature motion gives the character an aura of frustration or excitement. When you see R2-D2 in the first trilogy (Episodes IV through VI), the two-legged unit is usually Kenny Baker in costume, while the three-legged robots are invariably animatronic.

COMPUTER GRAPHICS

Physical puppets give the actors something to work with—in effect, to act with—but they do have certain limitations when it comes to gravity. Computer graphics (CG) characters, on the other hand, can run, jump, and even fly. Just as radio-controlled puppets leave behind the rods or cables of traditional puppets, CG robots are unconcerned with such sticky issues as self-balancing, or spindly limbs that wouldn't work mechanically in the real world. CG robots aren't bound by the laws of physics—just plausibility. The destroyer droids of Episodes I through III , for instance, have a three-legged configuration that yields a difficult, if not impossible, gait to replicate in real life. Only through the skill of the digital animators do the movie robots manage a realistic walking sequence on screen. Because they exist in a virtual world, CG characters don't have to worry about gravity, or supporting too much weight on spindly limbs, or breaking their ankles when they walk.

In the case of a flying character, such as R2-D2 in the droid factory of *Star Wars:* Episode II *Attack of the Clones*, the use of computer graphics does away with the need to design a complicated and potentially dangerous flying-wire rig for both puppet and camera. Instead, a weightless digital character in a digital environment offers up a flawless performance.

CG robots possess another key advantage: They allow for incredible flexibility in the postproduction process. They can be duplicated easily, into an army of thousands if need be, and their performances can be tweaked long after principal photography with the actors has been completed. Late in the filmmaking process, the director can still change CG designs, configurations—even the paint scheme. The price you pay for all this is that the actors often find themselves waving their lightsabers at imaginary objects and talking to nonexistent characters, which can impact their performance.

CREATING A PERSONALITY

Even if a movie robot moves in a completely realistic way, the audience won't relate to it unless it has a personality. Since many droids have no lips, no eyebrows, and no way to express emotion through their eyes, it's up to the skill of the puppeteer, actor, or digital animator to bring these characters to life. Depending upon character complexity, this can require more than one actor or operator.

In the case of C-3PO, the unique body language and voice of Anthony Daniels succeed in turning that golden robot shell into an endearingly fussy character. We identify with his feelings through such acting details as the way he tilts his head, or kicks R2-D2 in frustration. R2-D2, on the other hand, primarily uses sound rather than movement to develop personality. A series of beeps, clicks, and whistles enables the robot to communicate with other characters, although in reality the puppet makes none of these noises. They're added by Ben Burtt during the sound design process, long after the film sequence has been shot.

Action stories often call upon their characters to run and jump, shoot guns, and sometimes even fly. However, movie robots don't have to master all these skills; in the most basic sense, they have to do only enough to fool the camera. All that counts in the end is what you actually see on the screen, and whether or not you can suspend disbelief long enough to enjoy it. The many techniques employed in *Star Wars* form the bag of tricks used by visual effects artists to fool the senses. Some characters, such as R2-D2, use variations of all these tricks, in different combinations at different times. When we're successful, the transitions are seamless. And if we've done our jobs well in creating the characters, the audience will embrace them with open arms and make them a part of pop culture—as they have with R2-D2 and C-3PO.

Destroyer droids known as droidekas have legs, wheels, deflector shield generators, and twin blasters; all these design features are in aid of one goal: the total annihilation of their enemies. When droidekas roll onto the scene, audiences know that their Jedi opponents have their work cut out for them.

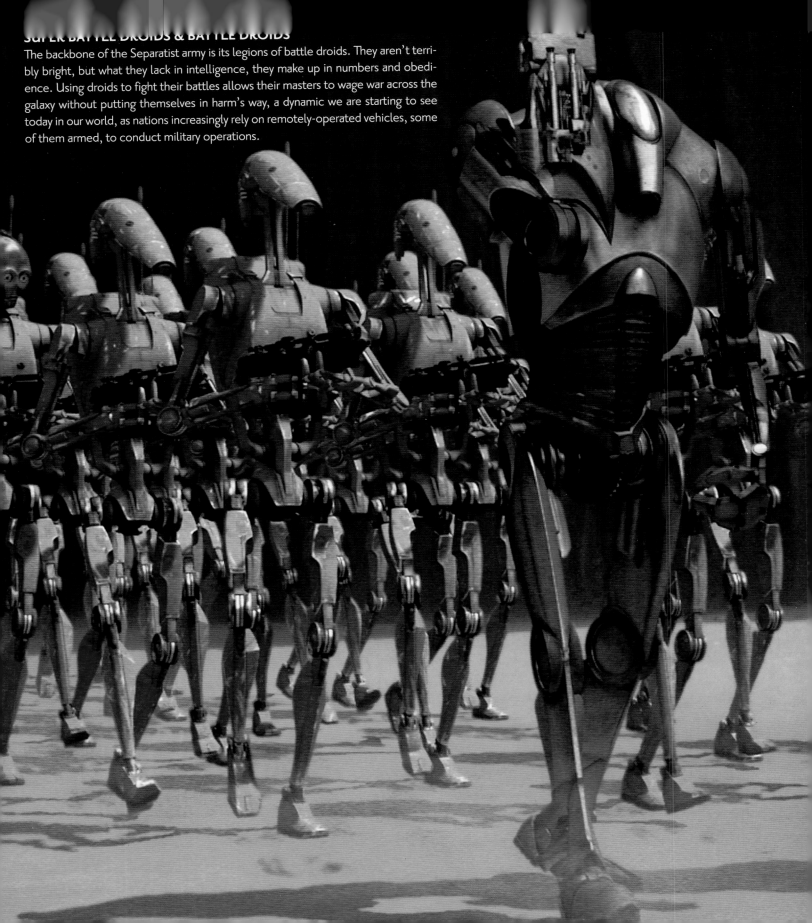

SUPER BATTLE DROIDS & BATTLE DROIDS

The backbone of the Separatist army is its legions of battle droids. They aren't terribly bright, but what they lack in intelligence, they make up in numbers and obedience. Using droids to fight their battles allows their masters to wage war across the galaxy without putting themselves in harm's way, a dynamic we are starting to see today in our world, as nations increasingly rely on remotely-operated vehicles, some of them armed, to conduct military operations.

STAR WARS DROIDS AND CYBORGS

A GALLERY

IMPERIAL PROBE DROID

Of all the droids in the *Star Wars* saga, the probe droid is perhaps the most recognizable to us, in that its mission is identical to that of the robots we ourselves
send out to explore the other worlds of our solar system. The Empire dispatched
hundreds of probe droids to scour remote areas of the *Star Wars* galaxy, looking
for signs of Rebel activity. On the icy planet of Hoth, this droid discovered the
Rebel base, only to be discovered in turn—and destroyed—by Han Solo.

ROBOTS AND PEOPLE

SEEKER DROIDS OF THE SITH

Spies are everywhere, even in remote corners of the *Star Wars* galaxy. Darth Maul uses the seeker droid, left, to track Queen Amidala and her Jedi protectors in Episode I; clone troopers search for Obi-Wan with versions of a seeker droid, right, after receiving infamous Order 66 in Episode III.

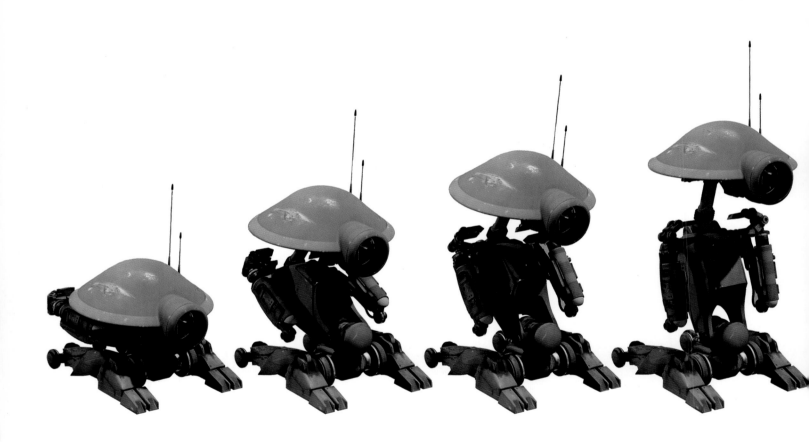

PIT DROID

The pit droids of Episode I *The Phantom Menace* are some of the most lovable minor characters in the *Star Wars* saga. Basically, they serve as portable, walking mechanics who scurry and bustle about, keeping the Podracers primed for speed.

ROBOTS AND PEOPLE

Bop a pit droid on the head, and presto! It unfolds into a small, bipedal robot. This illustration of how a pit droid extends from its compact, folded position to a fully upright stance shows the level of detail that goes into the design of even minor characters.

GENERAL GRIEVOUS

When we meet General Grievous in Episode III, he is entirely mechanical except for his eyes, brain, and some internal organs. This not entirely successful blending of natural and artificial is what gives him his added menace. Grievous's wheezing and persistent cough indicate that Darth Sidious and Count Dooku were not entirely successful in their attempt to make Grievous into a superbeing.

ROBOTS AND PEOPLE

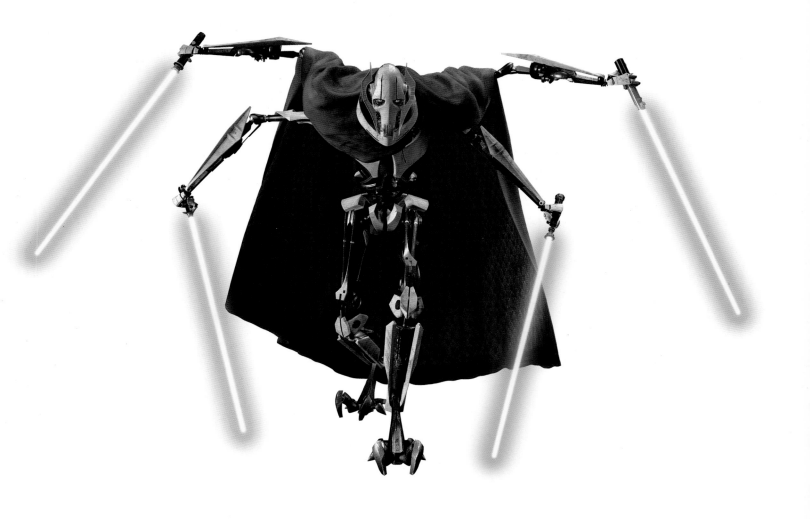

General Grievous is full of surprises. Not only has he been trained in the Jedi arts by none other than Count Dooku, he has also defeated several Jedi in battle. He keeps their lightsabers as trophies of his victories. Although he appears to be humanoid, Grievous possesses some unique functions. Most disturbing for the Jedi who face him is his ability to split each of his powerful arms. In a matter of seconds, he can change his two fully functional arms into four, each capable of wielding a lightsaber.

C-3PO

To look at him, you would never guess that C-3PO was assembled from scraps by a child. Protocol droids like C-3PO play a vital function, allowing the countless races that inhabit the *Star Wars* galaxy to understand one another. Tall, elegantly designed, and knowledgeable about humans, he is the polar opposite of R2-D2. In spite of his vast knowledge, however, he is constantly baffled and confused by the emotions of the humans he serves.

Astromech droids like R2-D2 are designed to work aboard starships. Their plethora of tools, probes, and sensors combine maximum functionality in as compact a design as possible. Since his primary function is to service vehicles, R2-D2 can speak only in the language of droids. Despite his limited vocabulary, everyone seems to understand R2-D2. And though his specialty is starships, he seems to have a deeper understanding of people than his counterpart C-3PO.

DIGITAL FLESH AND BONES
COMPUTER-GENERATED CHARACTERS

ROB COLEMAN

ROB COLEMAN Animation Director, Industrial Light & Magic. As animation director for *Star Wars* Episodes I-III, Coleman helped create such characters as General Grievous, the boga lizard, and vast armies of Gungans and battle droids—and has brought Yoda to digital life.

WE IN THE ANIMATION BUSINESS HAVE BEEN CREATING LIFE FOR JUST OVER A HUNDRED YEARS. We employ the alchemy of moving images—projecting 24 still pictures per second on a screen. By slightly altering the position of a limb or the expression of a face on each successive frame, we can create movement. If we are especially clever we can even convince the audience that our animated creature is alive and thinking. For years this has been referred to as the "Illusion of Life."

Computer animation represents the blending of art and science. Our creatures live their "lives" on the cinema screen, or on the DVD in your home, but they always begin as the spark of an idea in someone's mind, and then evolve into a quick sketch on a piece of paper. In the *Star Wars* films a character may go through hundreds of refinements—changes to its facial design, physique, and color—before it receives approval to be in our movies. As animators, we are interested in every aspect of the character's physical and emotional state.

During the design phase, and later in the animation phase, we continuously discuss aspects of each character. Through briefings with George Lucas, we may know only the planet the creature lives on, and no other details. Rarely at this early stage do we have a script or dialogue to refer to. It is up to us to make our creations believable.

The art directors and character designers, working at Skywalker Ranch, rely on environmental clues to help guide them in the development of a new species. They must ask themselves a series of questions: Does this creature live in the water, or the air? Is its habitat hot or very cold? How has it adapted to its environment? Does this particular character actually come from the world it is located on? By answering these basic questions during the design process, animators can use their knowledge of living on our own blue planet as a guide.

The Earth is a wondrous place—think of the variety of environs we have on our little planet, from the ice fields of the Arctic to the burning sands of the deserts, the lush jungles of the equatorial band, and the magical undersea worlds of the oceans. Each of these unique worlds has animals and plants that have evolved in order to live there, and each one of them has helped guide the *Star Wars* designers and animators to create the variety of life we find in the *Star Wars* galaxy. Evolution on the Earth has been working on creature design for millions of years; we would be foolish to ignore it. (Early on in the development of the Wookiee

Princess Leia called the Wookiee Chewbacca a "walking carpet," though she grew to appreciate his loyalty and fighting spirit. Wookiee chieftain Merumeru, played by Australian basketball player Axel Dench, strikes a warrior's pose as he leads his forces into battle. As fierce as they are furry, Wookiees help Yoda escape from their planet when his troops betray him.

Bad to the "bone": This special-ized droid, an IG-100-series MagnaGuard serving General Grievous, radiates threatening intent. The droid fights on even after Obi-Wan beheads it in *Revenge of the Sith*.

ROBOTS AND PEOPLE

environment—
Kashyyyk—several anima-
tors expressed concern because they
had seen paintings of this tropical planet with its
majestic Wookiees. They thought it strange that these
furry creatures would live in such a hot environment,
until they remembered that orangutans live in
Borneo and Sumatra, very close to the equator.)

Humans and animals move, and joints enable
this movement. We start the creature design with the
skeleton; does it have two legs, four legs, or more? Our computer characters are built with bones and joints. Each joint—a
wrist or a knee, for example—can rotate just as in our own human
bodies. Your wrist can bend only so far before it breaks, and this is also true
for our animated creatures.

Animators interest themselves in kinematics, the physics of a body in motion. To animate, we pose these digital bodies by rotating their joints. We can position the legs, the
back, the arms, and the neck any way we want. Think of these poses as snapshots, one
instant in time. If we create multiple poses and separate them by a few frames or seconds,
we create movement. Imagine someone throwing a ball, with the first position when the ball
is in a character's hand behind his head, and the second position when his arm is front of
him just as the ball leaves his fingers. If twelve frames separate those two poses, then it will
take half a second for that character to throw the ball. That is animation.

The size of a creature plays an important role in its motion because size determines how
quickly a creature can move. A mouse moves differently from an elephant, and likewise
Yoda and Chewbacca have different movements. Although Yoda and Chewbacca both stand
on two legs, their leg lengths necessitate different walks. After bearing in mind size, an animator will delve deeper into the character's locomotion and consider its age and strength to
help determine how it would move. We always have to remind ourselves that the characters
have mass. Their weight affects movement and it also affects inertia. It comes down to the
laws of physics—we have to animate the effect of gravity on our characters.

Yoda is a central character in the *Star Wars* saga. Originally a puppet performed by Frank
Oz in *The Empire Strikes Back*, he had to evolve into a digitally animated creature in the prequel
trilogy so that he could fulfill George Lucas's vision of him as a master swordsman. The animators at Industrial Light & Magic (ILM) carefully studied the puppet actions of the frail old
Jedi and attempted to distill the essence of his body's movements. Although we could free him

A whirling dervish in his dueling
scenes in Episodes II-III, Yoda
first appeared as a puppet operated by Frank Oz in Episode V,
The Empire Strikes Back. Not
until computer animation
caught up with Lucas's vision
could Yoda stand—or, when
need be, leap—and fight.

Initial design work for characters like Sebulba was done in traditional media, like pen and ink and physical models. Once Lucas approved the design, work could begin on creating the CG version seen on film.

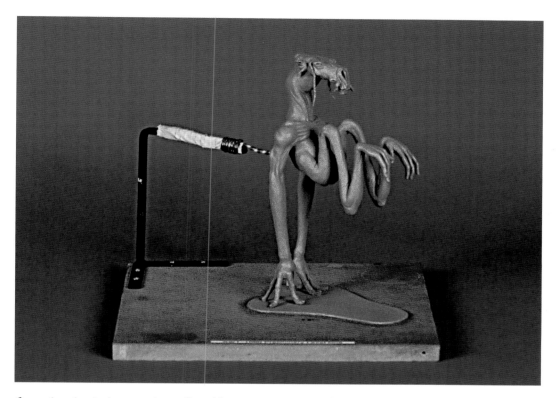

from the physical constraints of a rubber puppet, we still knew it was of the utmost importance to remain true to who Yoda was. We analyzed his range of facial movement, his walk, his skin and clothes—all of which held clues. Like forensic scientists, we deconstructed him down to his skeleton so that we could understand how and why he moved the way he did. In early tests we found that if we left out details like the tips of his ears wiggling, he didn't look like himself— he looked too young. Interestingly, Frank Oz told us later that he never intended for the puppet ears to quiver; his hand actions inside Yoda's puppet skull simply transferred out to the rubber ears. As Oz said, the ILM animators even copied his "mistakes." For instance, we found that if we gave Yoda too much articulation in his mouth he didn't look right. The original puppet could not round its lips to form the "o" and "u" vowel sounds, so we needed to be restrained with the digital version, and not stray too far from the established range of puppet actions.

Digital technology and the skills of computer graphics practitioners have developed dramatically over the past few years. Designers can now render lifelike skin, hair, and clothes in the computer. We have succeeded in writing software programs that emulate the way light reflects and refracts off real surfaces. By applying our understanding of digital illumination, we can create photo-realistic digital characters. This is especially challenging when you consider that our digital creations share the screen with live-action counterparts—that is, real people. The audience has the opportunity to assess our digital renditions as they appear next to the real thing; if, for example, the cloth isn't exactly right, the illusion will fail.

In the early stages of script development, concept artists created dozens of sketches, paintings, and models for every alien, vehicle, and prop to be used. George Lucas would mark his favorites with a rare "FABULOSO" stamp.

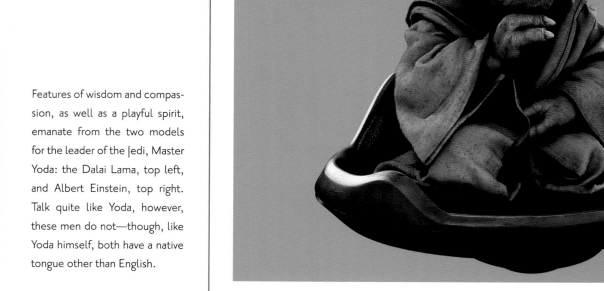

Features of wisdom and compassion, as well as a playful spirit, emanate from the two models for the leader of the Jedi, Master Yoda: the Dalai Lama, top left, and Albert Einstein, top right. Talk quite like Yoda, however, these men do not—though, like Yoda himself, both have a native tongue other than English.

The greatest challenge we faced centered on the acting aspect of the animated performance. It was of the utmost importance that we present Yoda as a thinking, breathing character. Subtle actions, like his eye blinks or the inhalation of breath, go a long way toward creating the illusion that he, or any other animated character, is alive.

In a sense, animators work like actors behind the scenes. We spend our lives studying human motion and emotions, considering how we might apply this knowledge to our characters. We carefully observe how people act, how they talk with their hands, how their faces change while they think about something or when they are upset. Really, all human beings are finely tuned observers. We look at faces from the day we are born and we learn how to read emotions—we know the physiology of the face almost at a subconscious level.

Animators routinely must tackle the complexities of human expression and emotion. Imagine the convoluted concept of an animated character saying one thing, but meaning another. Although we witness this in our daily lives, it represents a very difficult animation challenge to overcome.

We know that the eyes are the windows to the soul, and that is absolutely true for animation. Delicate changes in eye direction or a subtle tightening of the eyelids can help achieve that life behind the eyes. As for Yoda, the Dalai Lama and Albert Einstein have served as inspiration for his eyes—both very wise men, with a joy for life and laughter; we can read that in the wrinkles.

Vocal performances and changing intonations can affect everything, from how characters move their hands to their facial expressions. The syntax of the speech is also an important consideration. We have always assumed that Yoda speaks many different languages and that English is not his first language. As such, he constructs his thoughts in a different order from a native English speaker. He has developed a particular cadence to his speech, which has become an essential characteristic of his persona.

Junior animators always prefer scenes with lots of dialogue; to them, if the character is talking, he is acting. But the senior animators know that many times the nonspeaking moments hold the real acting gems: The quiet, nonverbal shots can present the subtlest examples of acting, and the look Yoda gives someone can tell you more about what he thinks than his spoken words. The real magic—the true magic—occurs when the character isn't speaking at all, but there is "life" in his eyes.

It was technically impossible to achieve a digital Yoda until the late 1990s. George Lucas had been waiting for the technology and artistry to catch up to his imagination. Although he had been very pleased with the puppet Yoda in the 1980s, he had longed to be free of the animatronics and cables. Lucas had always wanted the freedom to have Yoda run and fight.

Computer animation and digital visual effects have allowed Lucas to fill his prequel *Star Wars* films with the creatures of his imagination. Whether they are tiny characters, flying characters, or enormous characters, ILM has been able to transfer his visions into photorealistic digital reality—the illusion of living flesh and bone.

ASTRONAUTS, ROBONAUTS, AND THE MOON

Illah R. Nourbakhsh

ILLAH R. NOURBAKHSH, Robotics Group Leader at NASA/Ames Research Center, and Associate Professor of Robotics, Carnegie Mellon University. Dr. Nourbakhsh has wide-ranging interests within robotics, including human-robot collaboration and the educational uses of robots.

Called the first true spaceship because it flew only in the blackness of space, the Apollo lunar module carried astronauts to and from the moon.

WE ARE RETURNING TO THE MOON. THE LAST ASTRONAUTS WHO LANDED ON TAURUS-Littrow Valley, in 1972, would scarcely have believed that for more than thirty years there would be no new human footprints in the lunar soil. In 2005 NASA initiated a plan for going back to the moon, and it is likely that, within 20 years, the descendant of the Internet will be abuzz with live video of humans taking the 21st century's first steps on "terra luna." Although there will be similarities to the earlier Apollo missions—in the spirit of discovery and the wonderment of an unexplored world—the technology used in the next moon landing will be completely different from that of the earlier missions, and it will involve more than just people.

If all goes according to plan, this mission will inaugurate our first permanent settlement on another world, and it will probably be built almost entirely by robots. With luck, the next astronauts to reach the moon will step off the lunar lander, walk down a ready-made, pure lunar-concrete ramp to a pristine underground habitat, unlock the door, and enter a pressurized room where they can take off their helmets, hand their luggage to robotic bellboys, and celebrate their new abode. Like *Star Wars* characters, we will go into space with our robots at our sides. Only they probably won't look much like C-3PO or R2-D2.

The Apollo missions catalyzed research and development projects that ultimately led to technological innovations and products we now use to improve our daily lives here on Earth—from the cordless drill to Teflon, water filters, and kidney dialysis machines. The next set of missions to the moon will require building and maintaining long-term lunar habitats before humans ever arrive, and this will demand a new level of research and development, especially in the field of robotics. We will have to deploy robots more complicated and capable than any yet built on Earth. In virtually every aspect of human planetary space exploration, robotics will play a significant role in making this exploration safer and more affordable. Just as with Apollo, these advancements will have a profound ripple effect, and by learning more about them, we can begin to imagine how space robotics will improve our own lives and those of generations to come. We may be designing robots to work on the moon, but if history is any indication, those robots will certainly be put to work here at home as well.

In 1972, astronaut John Young followed in Neil Armstrong's footsteps, making another giant leap for mankind on the moon. Future missions to the lunar landscape may well be manned by robots.

One of the curious things about *Star Wars* films is that you rarely see the characters in any kind of hostile natural environment, like a planet with a methane atmosphere or a gravity field three times that of Earth, or, for that matter, like space itself. In this respect, real space exploration is far more alien than in *Star Wars*. On Earth, in the environment that has shaped our evolution, we manage to survive pretty easily, all things considered. Our environment provides us with the air we breathe, the light that powers all plant life on Earth, and abundant food. Step into a slightly more alien environment, however, and our terrestrial hierarchy of needs—food, shelter, clothing—becomes more complicated, or, looking at it another way, more fundamental. Scuba divers know that one rule ranks above all others: *Always breathe.* Holding your breath during an underwater dive is incredibly dangerous; approach the surface under such conditions and your lungs will explode as water pressure decreases. We gain no intuition for this basic rule by living on Earth's topside; the undersea environment is so foreign to us that it imposes new and exotic demands on our behavior. If you were going to live underwater, your hierarchy of needs would then be: air, clothing, food, and shelter. The spacefarer, even more removed from normal life, has needs that differ even more greatly from those of earthbound humans.

In 1943, psychologist Abraham H. Maslow famously wrote "A Theory of Human Motivation," suggesting an architecture for the needs and resulting motivations that drive humans forward in commonly observed behaviors. His model suggests that the most primitive needs of humans are physiological: food, shelter, clothing. Meet these basic needs, and safety, social love, and esteem take center stage in driving human behavior. Maslow's "Hierarchy of Needs" can help inform the study of human behavior on Earth, but we need to formulate a new hierarchy to reach an understanding of what will drive human behavior, and human technology, in space.

At the most primitive level astronauts also have base physiological needs, although shelter moves to the top of the list in space. Venture outside the protective cover provided by Earth's atmosphere and damaging radiation rains down with regularity, courtesy of the sun; in such a setting, radiation shelter is far more critical even than food. The human body, leveraging the magic of organic chemistry, turns its food into human heat and human motion with a surprising efficiency and elegance that no robotic system can match. But while food continues to be a physiological need beyond the Earth, energy on its own will become a new, basic requirement for human life in space.

Energy must provide a stable, comfortable temperature within an astronaut's spacesuit; it must provide astronauts with an oxygen-rich breathing environment and filter out excesses of carbon dioxide. Take any spacesuit or spaceship, cut off its power supply, and a life-or-death emergency for the inhabitants will ensue. Electricity, a modern convenience of life on Earth, is as important as blood for humans on the moon. The spacefarer's mantra may well be: *Never, ever run out of power.*

Maslow claims that once a human's physiological needs are met, that person is motivated further by social needs—by interactions with his or her community. In space, meeting the physiological needs of astronauts imposes significant limitations that drive the next higher needs, which we call "reach." The radiation shelter and energy needs will lead humans on the moon to live in buried shelters, venturing out-of-doors on extra-vehicular activities (EVAs) rarely, and then only in preplanned adventures. Spontaneity—and, with it, the very human behavior of extemporaneously responding to changing conditions by changing plans—will be lost in a thicket of safety, energy, and cost constraints. Humans who must function under such circumstances will need an effective reach that extends well beyond their physical reach. They'll need, for example, to inspect the outside of the pressure vessel, to venture beyond the habitat for a core sample of lunar soil, to conduct important repairs on a lunar power plant. And they'll need to do all this without leaving the safety of their habitat.

Another subtlety within the realm of reach involves the need for experts to solve specialized problems. Suppose your terra-cotta roof has been damaged in a hailstorm. You find a local expert in ceramic roofs, who drives over to your house, climbs up to assess the damage, and then accomplishes the repair. But astronauts are not experts in all fields, and a lunar

Clones of the bounty hunter Jango Fett eat together while undergoing training. Maslow demonstrated that humans have far more complex needs and motivations than robots, and consequently require much more to stay alive and healthy during the rigors of space travel and exploration.

colony will need to make use of expertise that is back on Earth for both maintenance and exploration. Reach, in this case, will apply not only to the astronauts themselves but also to the experts on Earth, whose attention and advice will be necessary components of any lunar settlement, at least for the foreseeable future.

Robotics will play a critical role in supporting the astronaut's hierarchy of needs, from the physiological ingredients of radiation shelter and energy to the higher-level need of reach. There is, however, one additional physiological need that results from the role that robots will play, a role that puts them in close quarters with astronauts: robotic safety. In short, robots should never impose upon nearby humans the threat of physical harm. The design of robots that are truly safe in the context of shoulder-to-shoulder collaboration with humans represents a major departure from the fast, stiff-motored industrial assembly line of the Geonosis droid factory, where the droids are dead ringers for the kinds of industrial robots used in factories all over Earth today. The need for humans to walk down a hallway and pass a robot going the other way without an awkward meeting or, worse, collision, stretches the capabilities of current robotics.

The challenges are grand, and yet recent advances hold out hope that robots can play a meaningful role in creating human space shelter and extending human reach, all the while guaranteeing physiological human-and-robot safety. We don't yet have robotic partners as useful as the droids that kept the Lars family's moisture farm running on Tatooine, but we can begin to see how we might design and build robots that can coexist with humans in the real world outside the robotics lab.

RADIATION SHELTER

A safe, long-term lunar habitat must first and foremost provide shelter from radiation. The straightforward solution, a heavily fortified, lead-lined pressure vessel built on Earth and shipped to the moon, is completely impractical due to the logistical constraints of launching such a structure. In the world of space travel, the most expensive component of flight is weight itself. Adding several pounds to a payload results in an increased amount of fuel required to launch that heavier payload into space. But that fuel itself must make it part-way into space, which in turn demands increases in onboard fuel. An increase in fuel means larger fuel tanks, which ups the spacecraft's dry weight, further pushing the cycle of additional energy requirements. Before long, such a mission would require a brand-new launch vehicle, far larger than anything currently available, just to get off the ground. So minimizing launch weight is not just a goal, but an imperative.

To minimize costs, the moon might be colonized through multiple launchings of small, lightweight parts, rather than utilizing one huge launch. In spite of the need for radiation shelter on the moon, we have no practical means of sending shielding material, such as large plates of lead, from here to there. Luckily, the moon is the ultimate sandbox because its soil,

Payload: The load carried by a vehicle which is exclusive to its operation, such as passengers or instruments.

ROBOTS AND PEOPLE

Because DNA sustains damage from radiation exposure (arrow at top) and from heavy ion particles (arrow at bottom), humans require heavy shielding for protection, adding to spacecraft payload; with less need of protection, robots reduce weight constrictions and eliminate the risk to human life.

called regolith, just happens to act as a great radiation barrier when piled several feet high. Robotic concepts, using novel automated construction techniques, might hold the solution for exploiting the radiation-shielding property of lunar regolith.

Consider inkjet printers. With great accuracy and speed, these machines can deposit ink just where we want it, in patterns on pieces of flat paper. Now think of the lunar surface as the flat paper and regolith as the ink material. Could something like a mobile jet printer create structures on the Moon by scooping regolith, then spraying it in desired patterns? Could such a printer actually build structures larger than itself?

Certainly, the concept of creating an object through the incremental addition of layers is not new in itself. Printers with 3-D capabilities are becoming more and more common in many businesses that have to rapidly produce a single object. Plastic parts have been built this way for years, with layer-by-layer deposits forming a three-dimensional shape, in some cases even with moving parts inside. The challenge here lies in creating a lunar-deposition robot capable of building habitable structures, by collecting and using regolith as the expendable building agent. Imagine a large, wheeled construction robot rumbling around a lunar settlement site, piling up chunks of regolith like a mason building a wall. Or perhaps we'll have a more complicated scenario, where one robot does

NASA has ambitious plans for returning to the moon in the 21st century. One option under consideration involves sending specialized robots ahead of humans. These builder robots would use lunar soil to make a kind of cement, from which they would fabricate a moon base, all before any astronauts even arrive.

nothing but create blocks of lunar regolith which other robots dutifully carry off and assemble into larger structures.

Good news on this front: Regolith not only shields against radiation, but research has also established that it can be sintered into a strong, solid (2,000 PSI), concrete-like substance. That means the moon itself will provide us with the materials to build driveways, walls, and ramps on its surface. Of course, we'll need to make terrestrial prototypes first, so do not be surprised by earthbound experiments: Large foam lawn ornaments and doghouses, courtesy of giant printing robots, just may be coming soon to a neighborhood near you!

Astronauts looking to spend significant time on the lunar surface will need mobility as well as shelter. No doubt they'll be conducting science and prospecting operations, and in any case, staying in one place is likely to get old after a few weeks. What if the entire lunar habitat were mobile, so that a single habitat could provide long-term access to a number of lunar sites? Suppose that a habitat consisted of a number of modular pieces: We could fly these modules to the moon on one of many launch opportunities, and have them land reasonably close to one another at a lunar target location. Next, the modules would rectify themselves, establish contact with the other modules, agree on a meeting point, and go there, perhaps using a combination of wheels and legs. They would then attach to one another, test their pressure seal, and finally, as a single unit, the habitat would bury itself under just the right amount of regolith for radiation shielding.

One advantage of this is that the modules could later decouple, dig themselves out, and travel to a new lunar destination to start all over again: a self-propelled, modular, lunar RV. For roboticists on Earth, this represents a fantastically challenging assignment, involving the design of multiple medium-sized robots, also known as hab modules, with built-in features ensuring mobility, communication, digging, and coupling capabilities. Robotic capacities have witnessed significant advances in recent years, and the integration of these advances, even to the point of achieving mobile modular habitats on the moon, is definitely possible.

ROBOTIC SAFETY

The issue of safety in factories using manipulation robotics has long been built around a simple rule of separation. In other words, put a cage around the robot and insure that, when the robot is powered up, the cage doors are locked to keep all the humans out. This does indeed prevent humans and robots from touching, and that certainly protects the humans from the unforgiving, strong motors of industrial robots. But in the close quarters of space travel, lunar habitat construction, and lunar exploration, we expect robots and humans to work together successfully, often shoulder to shoulder, without endangering lives. Consider C-3PO. He very rarely touches his human masters, and then only with some very gentle gesture like a tap. This may be a by-product of Anthony Daniels' conception of how C-3PO should act, but that kind of programming would be a vital component of any robot designed to work closely with people—keep your distance, and be very careful around humans, because they're soft and bruise easily.

How do we humans maintain physical safety among ourselves—do we simply avoid human-to-human contact at work? Not at all. Observe interactions on a subway or bus. Humans constantly come in contact with one another—shoulder against shoulder, foot against foot. Even so, disaster very rarely ensues, because humans are designed for safe contact. Touch sensors quickly make our bodies aware of potential collisions, and even

Robotic pit droids make last-minute preparations for pod racing on Tatooine. A great challenge of real-world robotics lies in designing robots that can work alongside humans.

ROBOTS AND PEOPLE

Mars landers will need to perform a series of high-risk tasks autonomously, such as locating their landing site and correctly deploying their payload; the robots they carry would need to function both autonomously and cooperatively—unlike pit droids, opposite, who frequently get in each other's way.

when collisions do occur, they're harmless most of the time. Our muscles provide power without stiffness, so that casual blows are usually deflected and impact energy absorbed and dissipated.

In industrial applications, robots continue to use strong but stiff motors and joints, largely because human contact is simply not necessary for the robots to do their jobs: arc-welding car frames, painting curved surfaces, stuffing electronics boards with discrete electronic elements, and the like. The concept of cooperation between humans and robots on the moon has the potential to inspire a small revolution. If we can design and build a robot to coexist with humans on the moon, then such a robot should certainly be up to the challenge of moving down a crowded sidewalk or navigating the obstacle-strewn rooms of a house.

Robots will give tools to humans, help lift lunar modules, dig with humans, and help humans perform all manner of physical activities. These robots will have humans in their work space, and accordingly they must be safe. No motor can be inflexible, no joint can fail to bend in an emergency or due to a software failure. The focus of research here includes motor technology and control technology, and in both of these areas advances are being made yearly. Consider the recent humanoid robot successes from Honda, Sony, Toyota, and a number of universities in Japan. A Honda robotics manager, when asked to name the most difficult challenge facing engineers in that company's decades-long effort toward a bipedal robot, quipped that in fact the biggest challenge could be summed up in one word: motors.

When humans and robots can safely work side by side, we can expect to see a whole range of human-robot applications that, today, are restricted to movies like *Star Wars*. The secret will lie in the mechanisms: the motors and the control algorithms that will drive the safe robots of the future. These mechanisms will make today's humanoids seem like inelegant antiques, dating from a time when the *right* robot motors had not yet been invented.

Arguably the single greatest challenge in robotics is not actually robotics, per se, but rather the design of interaction between humans and robots. Robots, after all, serve as nothing but tools for the extension of human faculties, ranging from sight, touch, and sound, to cognitive skills such as inspection, maintenance, and assembly. Some space robots will face the challenge of togetherness with their human masters, but others will need to overcome a peculiar challenge that will catalyze new interaction modalities between humans and robots: distance. The moon is so far from Earth that, even at light-speed, communication between earthbound humans and lunar robots will take seconds at best.

Consider the problem of mating two connectors on the moon. One might argue that humans on Earth, with our plentiful supply of oxygen and food, are far cheaper to maintain than astronauts or, indeed, sophisticated robots. Let the robot be a straightforward remote-control machine, and let a human at mission control on Earth employ a joystick to move the connectors into position for a solid connection. Suppose that a human can "joystick" a robot instantaneously for such an operation. Providing haptic feedback to the human still pre-sents a research challenge—we want the human to feel the connectors fitting together as if the operator's hands were right there, on the moon. After all, if the connectors are slightly out of alignment, a firm robotic push may break the housing rather than seating the connector properly.

Now imagine a half-second delay in communication. The operator will be frustrated but will probably adapt, learning that what he or she feels right now is actually what the robot felt a split second ago. But what about a round-trip time of five or six seconds? The video image and the haptic feedback felt by the operator will always be out of date. The simple task of inserting one connector into another thus becomes an exercise in long periods of waiting. Now imagine this ground operator attempting to interact with a moving target, for example helping an astronaut on the moon who is performing an emergency habitat-

Haptic: Relating to or based on the sense of touch; characterized by a predilection for the sense of touch.

Apollo 9 astronaut David R. Scott stands in the open hatch of the mission command module. Extra-vehicular activity is both dangerous and strenuous, but provides the only means of working outside the confines of the craft. In the future, telepresence or robotics may reduce the need for such activities.

ROBOTS AND PEOPLE

repair operation. Six seconds of round-trip time lag will render the team inefficient to the point of being ineffective.

To extend human reach, we'll have to surmount the challenges of distance by enabling each human to interact with each robot *at the right level*. The habitat astronaut, with hands free from an inflexible EVA suit, should be able to directly tele-operate a lunar robot using a sophisticated telepresence cockpit that provides stereo vision and full tactile feedback, as if the astronaut were inside the robot's camera. The EVA astronaut, encumbered by a spacesuit and in the midst of high-priority tasks on the lunar surface, will need to gesture and speak to robots in order to make effective use of them. The ground-operations personnel will need to issue high-level commands, such as "mate that connector," without specifically directing all the real-time, fine-grained motions implied by the command. The robot must execute this high-level command with its own local perception and decision-making capabilities, and of course it must know when to stop and ask for human help.

It would be a tall order even to suggest that mere desktop computers be given interfaces that enable humans to interact with them at a range of levels, from high-level commands to direct operation. To do so on physically embodied robots, with complex visual, aural, and tactile perceptions, and the ability to push back and change the world, represents a mammoth challenge in the realization of successful space robotics. But when this interface problem is solved, the ramifications will extend well beyond the vacuum of space. Deep undersea robots, nuclear cleanup robots, autonomous aircraft, robotic surgical assistants, and caregiving robots for the elderly all need interaction systems that enable a diverse range of modalities, from voice and gesture to direct telepresence. Industrial and commercial applications already drive the advancement of robotic interface technologies, but the particularly hard problem of interaction over long distances, posed by the lunar robotic domain, forces the issue as never before. More buttons and more training, the easy answers of the past, will not suffice. Ironically, space exploration will force us to invent more *natural* ways for humans to interact with technological artifacts.

The challenges of a sustained human presence on the moon will undoubtedly launch significant advances in robotics. Initially those advances will be specific to the needs of space agencies such as NASA, since funding constraints dictate that research be tightly focused on the mission. But the parameters of space robotics for lunar habitation, astronaut-robot safety, and astronaut-robot interface are clearly fundamental to all future applications of robotics, and the progress we make in the field of space robotics will have visceral impact on robots back on Earth. In as little as three decades, safe robots may work shoulder to shoulder with humans, performing tasks as diverse as those undertaken by humans today—from housekeeping and caregiving to washing (robotic) cars … and maybe even participating in a pick-up game of human and robot soccer.

Telepresence: A system in which humans, through controls or sensors, operate distant machinery, such as robots located in a hostile environment; a virtual rather than tactile human-machine interface

EPILOGUE

ED RODLEY

IT HAS BEEN PROFOUNDLY SATISFYING TO WORK WITH SUCH A DIVERSE AND TALENTED group of brilliant thinkers, and to realize that people like the authors in this volume are some of the people creating the future. They embody many of the goals of *Star Wars: Where Science Meets Imagination*, especially in the joy they bring to their work, be it designing digital characters (I will confess now that Yoda was my favorite character in *Revenge of the Sith)* or sociable robots, or maglev mass transit systems. They have provided us with glimpses of what's going on in many different fields, but where will their work lead us in the coming century?

TRANSPORTATION

Who hasn't seen a *Star Wars* speeder or starfighter and not wanted to have the chance to ride in one, to freely move at will in all three dimensions with the ease of simply getting in your car? People in the *Star Wars* galaxy can—and do— go wherever they want. We aren't quite so fortunate in our transportation options. The vast majority of the galaxy lies beyond our direct reach and probably will for a long time. But even as we plan to return to the moon and venture to Mars, our technology has taken us to the farthest corners of the universe and the beginning of time. Through our telescopes and instruments, we have peered into the distant past and explored galaxies far, far away. It may be that we will never journey between stars as easily as Han Solo, but that does not mean the galaxy will remain a mystery to us.

One thing I find sobering to remember is that we actually live in age when we could send a person to one of our nearest neighboring star systems, if we chose to do so. Most of the technologies necessary to construct a rocket fast enough to get to Alpha Centauri in a lifetime already exist in some form or another. Nuclear technologies are quite robust and fairly mature. More speculative technologies like matter/antimatter engines may be further off, but they're not science fiction. Solar sails are being actively tested in orbit. Scientists at CERN have already successfully created antimatter, in tiny amounts. The science is solid. What remains to be mastered is the engineering to make the

Gleaming with menace, Darth Vader's infamous black helmet and mask reflect his dark purpose, and help the remnants of his body to function in an advanced cyborg state.

Obi-Wan Kenobi and Anakin (upper left) follow Padmé's would-be assassin in a harrowing car chase, Coruscant-style. High speeds and three dimensions add tweaks to the traffic on this completely urban planet.

technology practicable. Right now, any attempt to send people to the stars would be the most expensive undertaking in all of human history. Costly is still not the same as impossible. Going to the moon was expensive, very expensive. But go we did, and the technological benefits we accrued from that massive endeavor are still paying dividends.

On the terrestrial front, we can see that some of our most ubiquitous forms of transport are reaching the limits of their utility. The gasoline-powered automobile seems destined for the history books as oil stocks dwindle and fuel costs climb. What remains unclear is whether the car will evolve into a vehicle that uses a different fuel source, or be replaced by something we can't yet see. Will our future be full of cars or personal transporters or small flying vehicles? The FAA has already begun working on systems that could manage the comings and goings of countless small aircraft in the same way they currently direct air traffic. This infrastructure could support a whole new class of small aircraft making short trips to and fro. They wouldn't necessarily have to be traditional aircraft; they could be anything that flies and carries people. We just have to develop a commercially viable vehicle, and inventors all over the planet have been working on that for years. It's only a matter of time.

As the urban population of the planet has continued to climb, mass transportation has gone from being an amenity to a necessity. There will be too many people needing to get to and from work and home to support the personal automobile model we have developed. Mass transit systems seem to offer a viable, sustainable solution for the few billion people who will want to get around within cities. In countries with high urban populations like Japan and China, new technologies like maglev are being tested and deployed as a way to move people more quickly and quietly, and with significantly fewer environmental impacts, than traditional diesel or electric train systems. You could travel cross-country at speeds approaching those of commercial aircraft without ever leaving the ground by more than a couple of inches.

Advances in low-speed maglev have made it feasible to think about systems of personal rail vehicles that could travel on rail networks. You could signal for a rail vehicle, punch in your destination, and sit back while the vehicle whisked you to your destination at speeds equal to or greater than those you can manage in city driving today. And that's just one field of application for this technology. Maglev could power more than trains. You could launch aircraft and rockets using maglev, or power elevators, or make super-efficient motors that don't wear out.

ROBOTS

The droids of *Star Wars* are some of the richest characters in the saga. C-3PO and R2-D2 alone manage not only to convey tremendous emotional range, but also to have relationships with people that are complex and deep. It is clear to the audience that these droids possess that spark that separates them from mere machines. But they are also clearly not people and, as C-3PO often laments, their lot is an unhappy one. Droids can be bought, sold, smashed to bits, and in C-3PO's case, have their memories completely erased with no regard for how they might feel about it. How we decide to classify created beings like robots will have a profound influence on what our future looks like.

In Japan, where they take developing humanoid robots very seriously, researchers have already begun looking at all the ramifications of having robots coexist in our world. A simple example demonstrates how profound the impact will be. Imagine a robot walking down the street of your hometown, where nothing else has changed. Now imagine that robot unintentionally steps on someone's foot and breaks their toe. Whose fault is it and who makes restitution to the poor person who got stepped on? The robot's manufacturer, the programmer who wrote the program, or the owner of the robot? Or, maybe, the robot itself is to blame? If your pet injures someone, you are liable as the owner, but a robot is a different proposition altogether; the answer is not clear-cut and likely will only be decided after some robot actually steps on some person's toe. As so often happens with technologies, our adaptation to them is evolving, with many missteps and false starts. There is no right answer for how to use technology—only what we currently think is the best answer, and that answer will vary.

Let's return to the example above. A robot is walking down the street. We currently compel automobile owners to carry automotive insurance, so it seems plausible that we will likewise compel robot owners to insure themselves. How much insurance is enough? Who will decide that? What kind of insurance will there be? The insurance industry will have a new field of services to provide. Now, suppose the person whose toe was broken decides to sue. Whom will they sue? Will there be lawyers who specialize in robot law, the same way there are malpractice and personal injury lawyers? Law schools will have to develop courses in robot law, and doubtless some of them will specialize in it as a way of attracting prospective students. Now let us imagine that same robot walking down the street. Suppose some person stops the robot and knocks it down. Will it be a crime to interfere with somebody's robot? If so, will it be a misdemeanor or a felony? What if the robot is damaged? A new body of court cases will have to determine what the acceptable penalties will be. Imagine all the people whose lives and livelihoods will be influenced by the existence of robots, even if they have nothing to do with robot production, or never even own one themselves.

That's just a sample of potential ramifications. Robots will probably need to refuel themselves, which means some entrepreneur will provide that service, for a price. How will our robots pay for their fuel? Will they have access to our bank accounts? Will we have a special robot fuel card like the highway transponders that many commuters use to pay highway tolls? Follow the robot in this thought experiment through an entire day's worth of activities and you will see how much transformation will occur. Our future with robots will be more complex than the movies might lead us to believe.

Perhaps the biggest difference emerging between the robots of tomorrow and the droids of *Star Wars* will involve construction and operation. C-3PO and R2-D2 may retain their appeal to us as characters, but as robots, they're very old school. The hard metal and mechanistic designs are exactly the kinds of things robot designers are moving away from in their quests to make robots that are more sensitive, more adaptable, and more robust. I spent an evening talking with Hirochika Inoue, the éminence grise of Japanese humanoid robotics. When I asked about the greatest challenge facing roboticists in the 21st century, his answer was "softness." I didn't understand, so he explained that softness, which we take for granted, is something robots desperately need if they are to survive out in the big, bad world. When we fall, our softness cushions our vital organs, absorbs the blow, and lessens the impact.

When a metal and plastic robot falls, the shock goes through its entire body. Softness is a desirable trait. Inoue showed me a video of one of the latest humanoid robots in Japan, and its most noticeable feature was a "soft bottom." When it started to fall, it would curl up and fall backward onto that soft bottom—and then pick itself up, ready to return to work.

To survive in the world, animals have developed extremely delicate sensors, which they use constantly to understand what's going on around their bodies. If robots are going to survive in the same world, with all its perils and obstacles, they'll need senses at least as refined. Consider your own fingers. Pick up a pen or pencil and roll it around. See how much information the nerves in your fingertips can give you about the object: shape, hardness, surface detail. And that's just a tiny fraction of the surface area of your body, and only one sensory channel. At the same time you're moving that pen around, you're probably also looking at it, judging its position and size. Walk up and thump most any robot in existence today, and chances are the robot won't even know it. In order to protect themselves from harm, tomorrow's robots will probably be covered in sensors, the same way we are.

The most visible difference, though, will likely be what robots are made of—and it probably won't be metal and hard plastic. The intersection of bioengineering and robotics has already produced simple artificial muscle tissues that contract and expand when stimulated. Complicated three-dimensional shapes can be grown out of synthetic organic materials. Robots made of organic materials hold the potential to repair themselves, something we humans do all the time without even knowing it. Imagine if you had to see the doctor every time you cut yourself; that's where robotics currently stands. The robots of the future will need to be able to take better care of themselves. Combine all these factors and the robot you'll come up with won't look anything like a *Star Wars* droid, or any robot you've seen on Earth. Exactly what it will be, and what capabilities it will have, remains to be seen. I, for one, can't wait.

Has this book sparked your imagination? Then give in to it, and explore further! If nothing else, this project demonstrates how scientists, engineers, and filmmakers all draw inspiration from unlikely sources. To think more about technology and your own place in a technological world, dive into George Lucas's *Star Wars* films, and look at them with an eye to the technologies they depict. Ask questions, and see where the answers lead you. *Star Wars*, like all good art, is a mirror. Its power lies in its ability to show us ourselves and make us consider things we've taken for granted, or never thought about before. And the more we learn about the real world of technology and our relationship to it, the more empowered we will be to make the technology decisions awaiting us.

The fantasy is great fun to watch, but there is no substitute for the real thing. Wonders that defy description are out there, just waiting for you to discover them.

Droid duo C-3PO and R2-D2 secured their place in popular culture, leaving them, and us, to ponder what new wonders the future may hold.

For every person named here, there are probably two more I'm forgetting; I beg your indulgence and forgiveness.

The authors of the essays deserve all the credit for the success of this volume. They took substantial time out of busy schedules to write thoughtful, interesting pieces. My thanks especially to the folks from Lucasfilm and ILM, who managed to meet our deadlines while still trying to finish Episode III.

When Johnna Rizzo approached the Museum about collaborating with National Geographic Society on a companion volume, we were skeptical of our ability to add yet more tasks to already full plates. Her enthusiasm and belief won us over. Even though she left NGS before this book was completed, her mark on it is clear.

New endeavors need champions, and this book owes everything to Larry Bell's advocacy. He saw immediately that this was not just one more thing to get done, but the first of what we hope will be many collaborations. He never doubted that the book was a good idea and would be a success. When I had my own doubts, I could borrow some of his certainty.

I cannot truly acknowledge the debt I owe Jan Crocker. Aside from giving me my first real museum job, she has been a mentor and friend for over fifteen years. She juggled all the tasks and timelines, found extra help, and made the time available for me to write and edit. She told me that this would be a unique opportunity, and (in true mentor fashion) a way to stretch myself as a professional. As usual, she was right.

Whenever I was stretched too thin, I only had to think of Kathleen Holliday to feel a bit ashamed. As Director of Special Projects at Lucasfilm, Ltd., she was our conduit to Lucasfilm, a job she also performed for half a dozen other exhibition projects worldwide. Regardless of which continent she was on, Kathleen always answered my questions. "Dream big," she would say, and she meant it. That kind of support is beyond gold.

At National Geographic, Lisa Lytton, Chris Anderson, Margo Browning, Melissa Farris, and Dana Chivvis had the unenviable task of producing a complex book in too little time, with a completely green co-editor, yours truly. The hours we collectively spent on the phone discussing authors, deadlines, essays, photographs, and illustrations were exhausting, but well worth it in the end.

At Lucasfilm, Nancy Frisch could be counted on as the cheerful bearer of good news. Jonathan Rinzler read every word in the book and provided valuable editorial oversight and commentary from one of the most knowledgeable sources in the *Star Wars* galaxy.

At the Museum of Science, Susannah Marsh assisted me with research and much more. Her unflappability and natural good humor saved my sanity on many occasions. Carolyn Kirdahy and Andy Grilz helped make artifacts move safely to and from photographers.

Mike Ambrogi from DEKA Research & Development has been a constant source of inspiration, with a remarkable ability to make things happen. He gave me the opportunity to make my pitch to Dean Kamen in person, and, with Donna Tamzarian, shepherded Dean's essay to publication. At the MIT Media Lab, Polly Guggenheim was a lifeline. Answers to questions were never more than a call or email away. At CSAIL, Ann Whittaker and Ann McNamara performed similar yeoman duty. Last, but not least, I owe a great debt to everyone who worked on the exhibition *Star Wars: Where Science Meets Imagination,* and had to put up with me being busy with some mysterious book when I should have been working with them. Our overall success is due to their dedication, skill, and professionalism.

Museum of Science, Boston

ABOUT THE MUSEUM OF SCIENCE, BOSTON:

The mission of the Museum of Science, Boston is to stimulate interest in and further understanding of science and technology and their importance for individuals and society.

To accomplish this educational mission, the staff, volunteers, Overseers, and Trustees of the Museum are dedicated to attracting the broadest possible spectrum of participants, and involving them in activities, exhibits, and programs which will:

- encourage curiosity, questioning, and exploration;
- inform and educate;
- enhance a sense of personal achievement in learning;
- respect individual interests, backgrounds, and abilities; and
- promote life-long learning and informed and active citizenship.

All this is offered in the spirit that learning is exciting and fun at the Museum of Science. For more information, contact the Museum at (617)723-2500, TTY: (617)589-0417; or find out more at www.mos.org.

ABOUT LUCASFILM:

Lucasfilm Ltd. is one of the world's leading film and entertainment companies. Founded by George Lucas in 1971, it is a privately held, integrated entertainment company. In addition to its motion picture and television productions, the company's global businesses include Industrial Light & Magic and Skywalker Sound, LucasArts Entertainment, Lucas Licensing, Lucasfilm Animation, and Lucas Online. Lucasfilm's feature films have won 19 Oscars and its television projects have won 12 Emmy Awards.

INDEX

CONTRIBUTOR'S BIOGRAPHIES (in order of appearance)

ED RODLEY has been in the BMOS Exhibits department since 1987 and has worked on exhibitions as diverse as the Soviet space program, Leonardo da Vinci, and Egyptian archaeology. As a content developer, he is responsible for development of exhibitions, including research, interactive exhibits, and label writing.

ANTHONY DANIELS left the study of law to follow his vocation as an actor. His exceptional miming and vocal skills caused George Lucas to cast him as the droid C-3PO; Daniels appears in all six *Star Wars* films. He currently acts on stage and in television dramas, and lectures around the world on *Star Wars* and robotics.

LAWRENCE M. KRAUSS, Ambrose Swasey Professor of Physics, Professor of Astronomy, and Director of the Center for Education and Research in Cosmology and Astrophysics, Case Western Reserve University. Author of the books, *The Physics of Star Trek,* and *Hiding in the Mirror*, Prof. Krauss is an internationally known theoretical physicist with wide research interests, including the interface between elementary particle physics and cosmology, where his studies include the early universe, the nature of dark matter, general relativity, and neutrino astrophysics. Krauss contributes frequently to NPR and numerous magazines and newspapers, including the *New York Times*.

MARC G. MILLIS, Propulsion Physicist and Founding Architect of the Interstellar Flight Foundation. At NASA's Glenn Research Center in Cleveland, Ohio, Marc Millis has worked in operations engineering, research, and project management;

he has designed ion thrusters, cryogenic propellant equipment, and cockpit displays for free-fall aircraft. In 2004, he co-founded the Interstellar Flight Foundation, which aims to extend spaceflight research beyond government, industry, and academia.

SAM GUROL, Director of Maglev Systems, General Atomics. Dr. Gurol has 30 years of experience in the development of high-technology magnet systems. Specific projects—with defense, transportation, and research applications—include superconducting magnets for particle beam accelerators and nuclear fusion, magnetohydrodynamic submarine propulsion, rocket-powered magnetic levitation, and urban maglev. Dr. Gurol currently leads a team developing an urban maglev system for General Atomics.

HIROSHI NAKASHIMA Deputy Director General, Maglev System Development Division, Central Japan Railway Company. One of the early proponents of maglev technology, Dr. Nakashima has devoted his career to achieving maximum-speed rail travel without sacrificing safety.

ALEX JAEGER, Industrial Designer and Visual Effects Art Director, Industrial Light & Magic. Alex Jaeger works as a designer and art director for graphic novels, and as a freelance agent within the film industry, as well as for Industrial Light & Magic.

RODNEY BROOKS, Panasonic Professor of Robotics, and Director, Computer Science and Artificial Intelligence Laboratory, Massachusetts Institute of Technology. Co-founder and chief technical officer of iRobot Corp., Dr. Brooks has

held research or faculty positions at Carnegie Mellon and Stanford Universities prior to joining MIT in 1984. The author of numerous papers and books, Dr. Brooks is involved both in the engineering of intelligent robots that can operate in unstructured environments, and in the understanding of human intelligence through the construction of humanoid robots.

ROBERT NAEYE, Senior Editor, *Sky & Telescope*. Robert Naeye has published two books, *Through the Eyes of Hubble: The Birth, Life, and Violent Death of Stars* (1997), and *Signals from Space: The Chandra X-ray Observatory* (2000). In 2002, Naeye received an award for science journalism from the American Astronomical Society, and was named Professional Astronomer of the Year by the Astronomical Association of Northern California.

RICHARD M. SATAVA, MD, FACS, Professor of Surgery, University of Washington Medical Center, and Program Manager, Defense Advanced Research Projects Agency (DARPA). Following his surgical residency and research fellowship at the Mayo Clinic, Dr. Satava has been involved in military surgery as an active flight surgeon, as a MASH surgeon during the Grenada invasion, and as a hospital commander during Desert Storm, all the while continuing his clinical surgical practice. He has published on the subjects of surgery in space, telepresence surgery, and virtual reality surgical simulation.

CYNTHIA BREAZEAL, Associate Professor of Media Arts and Sciences, and Director, Media Lab Robotic Life Group, Massachusetts Institute of Technology. Dr. Breazeal's current research in the area of human-robot relations aims to create cooperative and capable robots that can work and learn in partnership with people. The author of *Designing Sociable Robots* (MIT Press, 2002), she has been featured in various national and international print and broadcast media, including NPR's *Morning Edition,* the *Washington Post,* and the *London Times.*

DEAN KAMEN President, DEKA Research and Development Corporation. As an inventor, physicist, and entrepreneur, Dean Kamen has dedicated his life to developing technologies that help people lead better lives. He holds more than 150 U.S. and foreign patents, many of them for innovative medical devices that have expanded the frontiers of health care. Among Kamen's proudest accomplishments is founding FIRST (For Inspiration and Recognition of Science and Technology), a multinational non-profit organization that aspires to transform culture by motivating the next generation to understand, use, and enjoy science, math, engineering, and technology through various innovative programs.

GRANT IMAHARA, Animatronics Engineer and Model Maker. An animatronics engineer and model maker for George Lucas's Industrial Light & Magic, Grant Imahara worked on such films as *Star Wars* (Episodes I, II, and III), *Terminator 3,* and *A.I.: Artificial Intelligence.* Imahara was one of R2-D2's few official operators, and is the author of *Kickin' Bot: An illustrated Guide to Building Combat Robots.*

ROB COLEMAN Animation Director, Industrial Light & Magic. Rob Coleman joined Industrial Light & Magic's team of animators in 1993 to work on the film *The Mask.* At ILM,

Coleman has focused on the combination of live action and animated characters; he is especially interested in raising the quality and acting performance of digital creatures and leads the effort to help ILM animators understand that they are actors too. He was animation director of *Star Wars'* prequel trilogy, Episodes I-III.

ILLAH R. NOURBAKHSH, Robotics Group Leader at NASA/Ames Research Center, and Associate Professor of Robotics, Carnegie Mellon University. A founder and chief scientist of Blue Pumpkin Software, Inc., as well as co-founder of the Toy Robots Initiative at the Robotics Institute, Dr. Nourbakhsh has worked in such fields as robotic exploration; educational and social robotics, including human-robot collaboration; and robot personality and locomotion. He recently authored the textbook, *Introduction to Autonomous Mobile Robots,* published by MIT Press.

ALL IMAGES COURTESY LUCASFILM LTD. EXCEPT THE FOLLOWING:

Cover: Michael Desmond/Boston Museum of Science

Pg. 14 Michael Desmond/Boston Museum of Science, 15 Michael Desmond/Boston Museum of Science, 18 Micheal Desmond/ Boston Museum of Science, 21 Micheal Desmond/Boston Museum of Science, 25 Laurent Guiraud, 29 Margaret Bourke-White/Time & Life Pictures/Getty Images, 31 Don Foley, 36-37 NASA/JPL/Cornell, 37 NASA/CXC/GSFC/U.Hwang et al., 38-39 Andy Dunaway, 40 NASA Center: Dryden Flight Research Center, 41 Scaled Composites, LLC, 42 Adrian Mann, 43 Adrian Mann, 44 NASA, 45 Don Foley, 46 NASA, 47 NASA STI (Scientific and Technical Information) Program, 50 (Bottom) CORBIS, 50 (Middle) DoD photo by Petty Officer 2nd Class William H. Ramsey, U.S. Navy, 56 Moller International, 59 (Bottom) NASA, 60-61 Don Foley, 64 Transrapid International, 65 Dick Post, 66 Dick Post, 81 Top and bottom images Jamie Rose, 88 Justin Guarigula, 93 Bettmann/CORBIS, 98-99 Cary Wolinksy, 100 Sam Ogden, 104 Cary Wolinksky, 106 DoD/Jeffrey S. Viano, 107 DoD/AIC Kurt Gibbons III, 108 (Bottom) Bettmann/CORBIS, 110 Zen Icknow/ CORBIS, 111 NASA/JPL, 112 Otto Bock, 113 Rick Friedman/Corbis,116 Courtesy of Medtronic, Inc., 118 Reuters/CORBIS, 119 Sony Electronics Inc., 122-123 Charles O'Rear/COR-BIS, 124 Cary Wolinksy 125 Cary Wolinsky, 126 Don Foley, 127 courtesy University of California-Berkeley, 130-131 Cary Wolinsky, 133 Sam Ogden, 134 Yoshikazu Tsuno/ AFP/Getty Images, 136-137 Cary Wolinksy, 141 Reuters/CORBIS, 143-143 Sam Ogden, 144 Kazuhiro Nogi/AFP/Getty Images, 147 Segway, 148 Cary Wolinksy, 149 Cary Wolinksy, 176 (Left) Louise Gubb/ CORBIS SABA, 176 (Right) Bettmann/ COR-BIS, 178 NASA, 179 NASA, 183 NASA Marshall Space Flight Center, 183 (Both images) NASA, 184 Don Foley, 188 Don Foley, 190 NASA

Founded in 1888, the National Geographic Society is one of the largest nonprofit scientific and educational organizations in the world. It reaches more than 285 million people worldwide each month through its official journal, NATIONAL GEOGRAPHIC, and its four other magazines; the National Geographic Channel; television documentaries; radio programs; films; books; videos and DVDs; maps; and interactive media. National Geographic has funded more than 8,000 scientific research projects and supports an education program combating geographic illiteracy.

For more information, please call 1-800-NGS LINE (647-5463) or write to the following address:

National Geographic Society
1145 17th Street N.W.
Washington, D.C. 20036-4688
U.S.A.

Log on to nationalgeographic.com; AOL Keyword: NatGeo.

Library of Congress Cataloging-in-Publication Data available upon request.

ISBN-10: 0-7922-6200-X
ISBN-13: 978-0-7922-6200-8
Printed in China

STAR WARS: WHERE SCIENCE MEETS IMAGINATION

EDITED BY ED RODLEY, THE MUSEUM OF SCIENCE, BOSTON

PUBLISHED BY THE NATIONAL GEOGRAPHIC SOCIETY

John M. Fahey, Jr., President and
                    Chief Executive Officer

Gilbert M. Grosvenor, Chairman of the Board

Nina D. Hoffman, Executive Vice President

PREPARED BY THE BOOK DIVISION

Kevin Mulroy, Senior Vice President and Publisher

Marianne R. Koszorus, Design Director

Kristin Hanneman, Illustrations Director

STAFF FOR THIS BOOK

Lisa Lytton, Senior Project Editor

Melissa Farris, Art Director

John C. Anderson, Illustrations Editor

Margo Browning, Text Editor

Dana Chivvis, Associate llustrations Editor

Rachel Sweeney, Illustrations Specialist

Cataldo Perrone, Design Assistant

Evan Thompson, Consultant

Gary Colbert, Production Director

MANUFACTURING AND QUALITY CONTROL

Christopher A. Liedel, Chief Financial Oμcer

Phillip L. Schlosser, Managing Director

John T. Dunn, Technical Director

LUCASFILM, LTD.

Amy Gary, Director of Publishing

Jonathan Rinzler, Senior Editor

Iain Morris, Art Editor

Kathleen Holliday, Director of Special Projects